MULBERRY SERICULTURE
PROBLEMS AND PROSPECTS

MULBERRY SERICULTURE
PROBLEMS AND PROSPECTS

Kamal Jaiswal
Sunil P. Trivedi
B.N. Pandey
A.K. Tripathi

A P H PUBLISHING CORPORATION
4435-36/7, ANSARI ROAD, DARYA GANJ
NEW DELHI-110 002

Published by
S.B. Nangia
A P H Publishing Corporation
4435-36/7, Ansari Road, Daryaganj
New Delhi 110002
Ph.: 23274050
E-mail : aphbooks@gmail.com

2025

Printed at
Balaji Offset
Navin Shahdara, Delhi 110032

ॐ सह नाववतु सह नौ भुनक्त
सह वीर्यं करवावहै।
तेजस्वि नावधीतमस्तु मा विद्विषावहै।।

[हे परमात्मन् आप हम दोनों गुरु एवं शिष्यों की साथ-साथ रक्षा करें, एवम् हम दोनों का साथ-साथ पालन-पोषण करें; हम दोनों साथ ही साथ बल को प्राप्त करें। हम दोनों की अध्ययन की हुई विद्या तेजपूर्ण हो, हम दोनों सर्वदा परस्पर स्नेहसूत्र से बंधे रहें, हमारे अन्दर कभी भी वैमनस्य न हो।।]

-उपनिषद्

PREFACE

Mulberry (*Morus* spp.) is an important commercial crop cultivated extensively both in temperate and tropical climatic conditions as a sole food plant for silkworm, *Bombyx mori*. Being perennial, deep rooted, high biomass producing foliage plant, mulberry gained importance in sericulture industry as a food resource for silk worms. Besides, other factors, no doubt, leaf quality of mulberry plays a pivotal role in up-keeping of silkworms and production of silk. This has generated an interest in Seri – biologists to develop a variety of improved strains of mulberry by continuous breeding efforts. Moreover, researches were also carried out to explore and develop disease resistant mulberry plants with superior leaf quality.

The present book entitled, **"Mulberry Sericulture: Problems and Prospects"**, has been painted on a large canvas with broad strokes comprising 19 chapters by experts and researchers delved in their respective fields covering the subject area. We appreciate the efforts of all the contributors whose prompt response gave us the necessary impetus to conquer the endeavor and publish the proceedings of **"18th All India Congress of Zoology and National Seminar on "Current Issues on Applied Zoology and Environmental Sciences with Special Reference to Eco-restoration & Management of Bioresources" (SCIAZE)** in the stipulated time period.

We express our heartfelt thanks to the patrons and organizers of 18th AICZ & SCIAZE. Sincere patronage, kind cooperation and generous help rendered by Hon'ble the Vice-chancellor and teaching and non-teaching staff of the University of Lucknow, Lucknow, for making the Congress and Seminar a success, are gratefully acknowledged. Our special thanks are due to the publisher of this volume, APH Publishing Corporation, for bringing out the book well in time.

CONTENTS

LIST OF CONTRIBUTORS

Mir Nisar Ahmad
Central Sericultural
Research and Training Institute (C.S.R.&T.I.),
Pampore-192121, J&K.

S.R. Ananthanarayana
Department of Sericulture,
Jnanabharathi Campus,
Bangalore University,
Bangalore-56, Karnataka.

M. Ramesh Babu
Andhra Pradesh Statc Scriculture Research and
Development Institute (APSSRDI), Kirikera - 515211,
Hindupur, A.P.

A.K. Bajpai
Central Sericultural Research and Training
Institute (C.S.R.&T.I.),
Berhampore-742101, W.B.

N. Balachandran
Cenral Sericultural Germplasm Resources Centre, Thally Road,
Hosur-635109, Tamilnadu.

M.M. Bhatt
Regional Sericultural Research Station,
Sahaspur, **Dehradun**- 248001, Uttrakhand

B.B. Bindroo
Regional Sericultural Research Station, **Miransahib**,
Jammu-181101, J&K.

T. Datta (Biswas)

Central Sericultural Research and Training Institute (C.S.R.&T.I.), **Berhampore**-742101, W.B.

Chandrashekharaiah

Andhra Pradesh State Sericulture Research and Development Institute (APSSRDI), Kirikera-515211, **Hindupur**, A.P.

T.P.S. Chauhan

Regional Sericultural Research Station, **Miransahib**, Jammu -181101, J&K.

S.K. Das

Central Sericultural Research and Training Institute (C.S.R.&T.I.), **Berhampore**-742101, W.B.

S.L. Dhar

Regional Sericultural Research Station, **Miransahib,** Jammu -181101, J&K.

Anil Dhar

Regional Sericultural Research Station, **Miransahib,** Jammu -181101, J&K.

Vadamalai Elangovan

Department of Applied Animal Science, Babasaheb Bhimrao Ambedkar University, Vidya Vihar, Raibareli Road, **Lucknow**-226025, U.P.

D. Gangopadyay

Central Sericultural Research and Training Institute (C.S.R.&T.I.), **Mysore**-570008, Karnataka.

C.R. Hegde

Department of Sericulture, Jnanabharathi Campus, Bangalore University, **Bangalore** - 560068, Karnataka

S.A. Hiremath

Cenral Sericultural Germplasm Resources Centre,
Thally Road, **Hosur** - 635109,
Tamilnadu.

C.K. Kamble

Central Sericultural Research and Training
Institute (C.S.R.&T.I.),
Mysore-570008, Karnataka.

M.A. Khan

Central Sericultural Research and Training
Institute (C.S.R.&T.I.),
Pampore-192121, J&K.

M.Z. Khan

Central Sericultural Research and Training
Institute (C.S.R.&T.I.),
Berhampore-742101, W.B.

R.K. Khatri

Zonal Silkworm Seed Organisation,
Central Silk Board Majra,
Dehradun-248171, Uttrakhand.

P.R. Koundinya

Cenral Sericultural Germplasm Resources Centre,
Thally Road, **Hosur**-635109, Tamilnadu.

Deepak Kumar

Regional Development Office,
Central Silk Board, 5th Floor,
"Vikas Deep", **Lucknow** - 226001, U.P.

M.V. Santha Kumar

Central Sericultural Research and Training
Institute (C.S.R.&T.I.),
Berhampore - 742101, W.B.

H. Lakshmi
Andhra Pradesh State Sericulture
Research and Development Institute,
Kirikera - 515211, **Hindupur**, A.P

R. Lochan
P-2, Basic Seed Farms,
Dehradun – 248001, Uttarakhand

K. Mandal
Central Sericultural Research and Training
Institute (C.S.R.&T.I.),
Berhampore - 742101, W.B.

P. Mitra
Central Sericultural Research and Training
Institute (C.S.R.&T.I.),
Berhampore - 742101, W.B.

B. Mohan
Cenral Sericultural Germplasm Resources Centre,
Thally Road, **Hosur** - 635109, Tamilnadu

S.K. Mukhopadhyay
Entomology Laboratory,
Central Sericultural Research and Training
Institute (C.S.R.&T.I.),
Berhampore - 742101, W.B.

Geetha N. Murthy
Central Silk Board, Ministry of Textiles,
Govt. of India, CSB Complex,
Bangalore-560068, Karnataka

M. Muthulakshmi
Central Sericultural Germplasm Resources Centre, Thally
Road, **Hosur**-635109, Tamilnadu

R. Nirupama
Central Sericultural Research and Training
Institute (C.S.R.&T.I.),
Mysore - 570008, Karnataka

H.S. Phaniraj
Central Silk Board, Ministry of Textiles,
Govt. of India, CSB Complex,
Bangalore - 560068, Karnataka

J. Prasad
Andhra Pradesh State Sericulture
Research and Development Institute (APSSRDI),
Kirikera-515211,
Hindupur, A.P.

S.M.H. Qadri
Cenral Sericultural Germplasm Resources Centre,
Thally Road, **Hosur**-635109, Tamilnadu

S.M. Qadir
P-4, Basic Seed Farm, Central Silk Board, **Manasbal**,
Kashmir, J&K.

S.K. Raina
Regional Sericultural Research Station, **Miransahib,**
Jammu - 181101, J&K

M. Venkateswara Rao
Central Sericultural Research and Training
Institute (C.S.R.&T.I.),
Sreerampura, **Mysore**-570008, Karnataka

A.K. Saha
Central Sericultural Research and Training
Institute (C.S.R.&T.I.),
Berhampore-742101, W.B.

B. Saratchandra
Central Silk Board, Ministry of Textiles,
Govt. of India, CSB Complex,
Bangalore-560068, Karnataka

J. Seetharamulu
Andhra Pradesh State Sericulture Research and Development
Institute (APSSRDI), Kirikera-515211,
Hindupur, A.P.

Abad A. Siddiqui
Central Sericultural Research and
Training Institute (C.S.R.&T.I.),
Pampore - 181101, J&K.

Ravindra Singh
Central Sericultural Research and Training
Institute (C.S.R.&T.I.),
Mysore-570008, Karnataka.

A.A. Siddiqui
Regional Development Office,
Central Silk Board, 5th Floor, "Vikas Deep",
Lucknow-226007, U.P.

G.B. Singh
Central Sericultural Research and Training Institute
(C.S.R.&T.I.), Sreerampura,
Mysore-570008, Karnataka

B.K. Singhal
Regional Sericultural Research Station, **Miransahib**,
Jammu - 181101, J&K

R.K. Sinha
Cenral Sericultural Germplasm Resources Centre,
Thally Road, **Hosur**-635109, Tamilnadu

D.P. Srivastava
Regional Development Office,
Central Silk Board, 5th Floor, "Vikas Deep",
Lucknow-226001, U.P.

Anshul Srivastava
Department of Applied Animal Science,
Babasaheb Bhimrao Ambedkar University,
Vidya Vihar, Raibareli Road,
Lucknow-226025, U.P.

Mukesh Tayal
Regional Sericultural Research Station,
Sahaspur, **Dehradun** - 248001, Uttrakhand

Ranjan Tiwari
Research Extension Centre, Central Silk Board,
Una - 174303, H.P.

T.S. Tsuchiya
J.I.C.A. Expert, Central Sericultural Research and Training Institute (C.S.R.&T.I.), Sreerampura,
Mysore-570008. Karnataka

S.K. Tyagi
Research Extension Centre, Central Silk Board,
Una - 174303, H.P.

T.M. Veeraiah
NSSO Head Quarters, Central Silk Board,
Bangalore-68, Karnataka

Meera Verma
Regional Development Office,
Central Silk Board, 5th Floor, "Vikas Deep",
Lucknow-226001, U.P.

NEW BREEDING APPROACHES IN THE DEVELOPMENT OF SILKWORM BREEDS OF THE MULBERRY SILKWORM, *BOMBYX MORI* L.

RAVINDRA SINGH, D. GANGOPADYAY, R. NIRUPAMA AND C.K. KAMBLE

ABSTRACT

Attempts have been made at Central Sericultural Research and Training Institute. Mysore. to develop superior silkworm breeds/hybrids of the mulberry silkworm. *Bombyx mori* L. through the application of artificial parthenogenesis and androgenesis. Response of different silkworm breeds towards parthenogenetic development has been studied. Among the different bivoltine and polyvoltine breeds, Japanese type bivoltine silkworm breeds showed pronounced parthenogenetic development. An increased tendency towards parthenogenesis was observed in hybrids obtained from a mother moth having a high tendency of parthenogenesis. Three bivoltine breeds *viz.*, DNB_1, DNB_6 and DNB_7 were developed. Evaluation of the developed breeds/hybrids was carried out through various statistical measures like analysis of combining ability, hybrid vigour and cocoon size uniformity. Studies showed more combining ability, hybrid vigour and cocoon size uniformity in the new hybrids. Induction of androgenesis was performed in different silkworms to select potential breeds. Polyvoltine hybrids showed higher androgenic development as compared to bivoltine hybrids. A breed with dominant cocoon colour gene was utilized as genetic marker to identify androgenic male individuals. Induction of androgenesis was performed by exposing the oviposited eggs to hot air (38 °C) for 200 min. Repeated backcrossing was adopted utilizing androgenic males to introgress homozygosity in the breeding lines.

Key words: Ameiotic Parthenogenesis, Androgenesis, *Bombyx mori* L.. Development, Parthenoclones, Selection.

Central Sericultural Research and Training Institute, Mysore- 570 008, India.

INTRODUCTION

In sexual reproduction, the fusion of ovum with a sperm leading to formation of a new zygote creates striking variability in the resulting organism. This variability favours the opportunity of natural and artificial selection. The mulberry silkworm, *Bombyx mori* L. is not an exception. Continuous domestication and selection has made the silkworm a diversified genotype (Tazima, 1964). Though, the conventional breeding approaches have remarkably increased the quantum of silk produced, continuous selection showed a decline of the targeted characters (Seidel and Brackett, 1981). Selection of desirable individuals based on phenotypic observation is sometimes not judicial. Therefore, breeders often opt to preserve the exact genotypical copy of the parent in the descendants (Strunnikov, 1983).

Some new breeding approaches like parthenogenesis and androgenesis would be useful in the development of superior silkworm breeds and hybrids with more viability, hybrid vigour, combining ability and less phenotypic variability (Dznealaidze and Tabliashvili, 1990; Takei *et al.,* 1990; Plugaru *et al.,* 1993; Strunnikov, 1995). Spontaneous parthenogenesis is rare in the silkworm, activation of excised unfertilized eggs at 46 °C hot water for 18 min after 12 h of extraction from ovarioles is effective in successful induction of artificial parthenogenesis (Astaurov, 1940). Androgenic development of a zygote by fusion of two sperm nuclei can facilitate to produce only male individual. Application of several activating agents like X-ray and gamma irradiation of mated female moths, hot air, hot water, CO_2 and super cooling of oviposited eggs at low temperature (-11 °C) were tried for induction of androgenic development in the silkworm (Whiting 1955; Tazima and Onuma, 1967; Ye *et al.*, 1989; Nagoya *et al.*, 1996). Attempts have also been made to isolate bisexual lines of the mulberry silkworm through the application of dispermic androgenesis (Xu *et al.*, 1997; Nacheva *et al.,* 1999). During the last few decades, Indian raw silk production reached a quantum jump and some promising hybrids like Jayalakshmi, Chamaraja and Krishnaraja are being largely exploited with the farmers. However, little information is available on the role of artificial parthenogenesis and androgenesis in the development of superior silkworm breeds/hybrids. The present study has been undertaken to explore the possibility of using new breeding strategies like artificial parthenogenesis and androgenesis in the development of superior hybrids of high viability, more hybrid vigour, combining ability and phenotypically uniform population.

MATERIALS AND METHODS

Resource materials and induction of artificial parthenogenesis: Females of 15 bivoltine breeds including 7 Chinese types *viz.*, CSR_2, CSR_2 (SL), CSR_3, CSR_{12}, CSR_{17} CSR_{18} and CSR_{27} and 8 Japanese types *viz.*, NB_4D_2, CSR_4, CSR_5, CSR_6, CSR_8 (SL), CSR_{16}, CSR_{19} and CSR_{26} were investigated to know their ability towards artificial parthenogenesis. Bivoltine silkworm breeds with high parthenogenetic ability were crossed and 9 bivoltine hybrids involving 5 Chinese types (*viz.*, $CSR_2 \times CSR_{17}$, $CSR_2 \times CSR_{27}$, $CSR_3 \times CSR_{17}$, $CSR_{17} \times CSR_{12}$, $CSR_{18} \times CSR_{12}$) and 4 Japanese types (*viz.* $NB_4D_2 \times CSR_4$, $CSR_6 \times CSR_{26}$, $CSR_{19} \times CSR_4$, $CSR_{19} \times CSR_6$) were prepared. Hybrid females with better response towards parthenogenesis were kept for future breeding programme.

Astaurov method (1940) was followed to induce ameiotic parthenogenesis in the unfertilized eggs. Eggs were extracted by rubbing through a muslin sieve, washed in running tap water, dried and kept in a cotton bag for 12 h at room temperature. Then they were dipped in a hot water bath at 46 °C for 18 min and abruptly cooled at 20 °C water bath for 10 min. After drying, eggs batches were put in a petridish and incubated at 15 °C and 80% RH for 3 - 5 days. Egg batches were soaked in hot HCl (Sp. Gr. 1.075) for 5 min. to terminate the egg diapause and rinsed in tap water

(20 °C) to eliminate acid traces before being dried. Care was taken to incubate the egg batches at normal temperature (25 ± 0.5 °C) and relatively higher humidity of 90-95% till larval hatching. Appearance of reddish-brown to dark-brown pigmentation in the serosa was considered as parthenogenetic development. The rate of parthenogenesis was estimated by counting pigmented eggs about 7 - 8 days after transfer from 15 to 25 °C. The ratio of reddish-brown/dark pigmented eggs and total number of eggs treated was expressed as percentage of parthenogenesis whereas the ratio of hatched larvae and pigmented eggs was considered as percentage of hatching. Data were recorded for number of pigmented eggs, number of non-pigmented eggs, number of larvae hatched, percentage parthenogenesis and hatching.

Resource materials and induction of androgenesis: Androgenic development in the hybrids of 3 polyvoltine, 2 polyvoltine × bivoltine, 2 bivoltine × polyvoltine and 3 bivoltine was studied utilizing 7 polyvoltine breeds *viz.*, BL_{68}, Mysore Princess, ND_7, Nistari, MAD, D_1, MR_1 (CSL) and 5 bivoltine silkworm breeds *viz.*, CSR_2, CSR_4, CSR_2(SL), CSR_{17} and NB_4D_2. Induction of androgenesis in the oviposited eggs was carried out in three different hot air treatments i) at 38UC for 200 min, ii) 40ÚC for 135 min and iii) 42ÚC for 210 min to standardize the procedure. Soon after treatment, the eggs were transferred to 15Ú C till the appearance of pigmentation in the serosa. Bivoltine and bivoltine × polyvoltine hybrids eggs were treated with hot hydrochloric acid to terminate the egg diapause. Incubation of eggs was done at 25ÚC till hatching. The ratio of dark bluish pigmented eggs and total number of eggs treated was expressed as percentage of androgenetic eggs.

Artificial parthenogenesis and silkworm breeding: Five breeding plans were initiated during the course of breeding for the development of homozygous breeds of the silkworm utilizing Chinese and Japanese type bivoltine silkworm breeds as breeding resource materials. Breeds with sex-limited characteristics were kept as genetic marker to identify the parthenogenetic female individuals. The Ps allele from the P multiallelic series responsible for the larval marking character was used as genetic marker. Six parthenogenetic lines namely, DNB_1 and DNB_2 of Chinese type (plain larvae; oval cocoons) and DNB_3, DNB_4, DNB_6 and DNB_7 having Japanese racial characteristics (marked larvae; dumbbell cocoons) were evolved. Characteristics of the newly developed breeds have been presented in Table 1. A breeding strategy to obtain homozygous breeds of the silkworm utilizing artificial parthenogenesis as a breeding tool has been schematically outlined in Fig. 1.1.

Androgenesis and silkworm breeding: Twenty six polyvoltine silkworm breeds *viz.*, BL_{23}, BL_{24}, BL_{61}, BL_{62}, BL_{65}, BL_{67}, BL_{68}, BL_{69}, 96A, 96C, 96B, ND_5, ND_7, NP_1, PM, P_2D_1, MY_1, D_1, GNP, Sarupat, Moria, Nistari, Kollegal Jawan, Kolar Gold, DNP_3 and DNP_5 were screened to shortlist superior breeds based on average evaluation indices as per Mano *et al.* (1993). Five polyvoltine silkworm breeds *viz.*, NP_1, ND_7, BL_{68}, DNP_3 and DNP_5 exhibiting higher evaluation index values were utilized as breeding resource materials. Nistari, a polyvoltine breed possessing dominant gene for golden yellow cocoon colour with marked larvae was utilized as genetic marker to identify androgenetic male individuals. Repeated backcrossing was adopted utilizing the androgenic males. A breeding strategy in the development of bisexual lines of the silkworm with androgenic origin has been presented in Fig. 1.2. Salient features of 5 bisexual polyvoltine lines with androgenic origin are given in Table 1.2.

Evaluation of silkworm hybrids: Six newly developed bivoltine breeding lines *viz.*, DNB_1, DNB_2, DNB_3, DNB_4, DNB_6 and DNB_7 were evaluated with 4 bivoltine breeds namely, CSR_2, CSR_4, CSR_{17} and NB_4D_2. Selection of the silkworm breeds/ hybrids was carried out through multiple traits evaluation index

Table 1.1. Salient features of developed bivoltine breeds/hybrids of the silkworm, *Bombyx mori* L.

Breed/Hybrid	Parents involved	Method adopted	Sex	Cocoon colour	Cocoon shape	Cocoon weight (g)	Cocoon shell weight (g)	Cocoon shell (%)
DNB_1	Parthenoclones of $CSR_{18} \times CSR_{12}$ backcrossed with CSR_2	Artificial parthenogenesis coupled with conventional breeding	Bisexual	White	Oval	1.759	0.373	21.22
DNB_6	$CSR_{19} \times CSR_4$	Artificial parthenogenesis	Parthenoclones with entirely females	Creamish-white	Dumbbell	1.973	0.380	19.26
DNB_7	Parthenoclones of $CSR_{19} \times CSR_4$ backcrossed with CSR_4	Artificial parthenogenesis coupled with conventional breeding	Parthenoclones with entirely females	White	Dumbbell	1.860	0.341	18.32
$DNB_1 \times CSR_4$	-	-	Bisexual	White	Intermediate between oval and dumbbell	1.825	0.397	21.75
$DNB_7 \times CSR_2$	-	-	Bisexual	White	Intermediate between oval and dumbbell	1.895	0.422	22.27

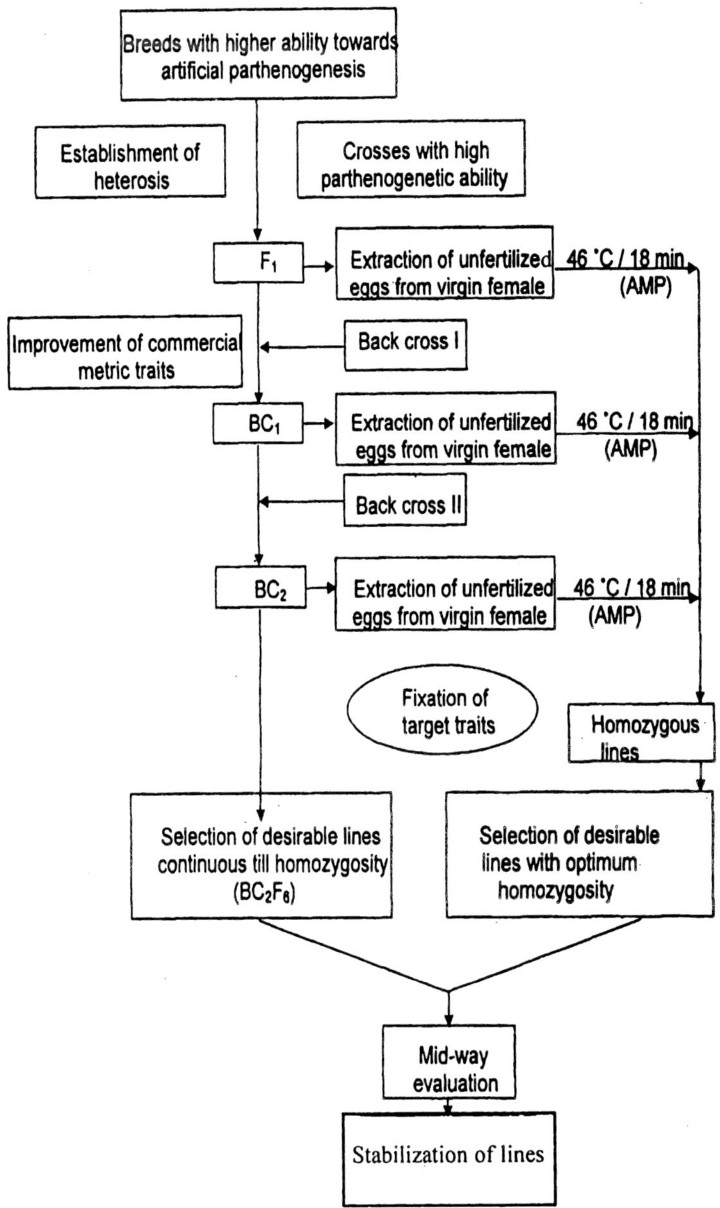

Fig. 1.1: A breeding strategy to obtain homozygous lines of the silkworm, *Bombyx mori* L.

method of Mano *et al.*, (1993). The breeds/hybrids showing greater average evaluation index value and evaluation index value for a particular character higher than 50 for more characters were identified as promising.

To understand the general combining ability of the parents and specific combining ability of the hybrids, the line × tester model of Kempthorne (1957) was followed. Newly developed 6 breeding lines and 4 popular bivoltine breeds were utilized as lines and testers respectively. For each line, tester and their hybrid, 3 replications were maintained. After third moult, 250 larvae were retained in each replication. Performance of the newly developed bivoltine was statistically compared with $CSR_2 \times CSR_4$ as control.

Heterosis over mid parent (MPH) and better parent (BPH) was calculated by using the following formulae:

$MPH = 100 \times (F_1 - MPV)/MPV$

$BPH = 100 \times (F_1 - BPV)/BPV$

Where, MPV = mid parent value

BPV = better parent value

To know the variability in cocoon size, one hundred F_1 hybrid cocoons were randomly picked up and three cocoon size variables *viz.*, cocoon length, cocoon width and length/width ratio were determined. Measurement of cocoon width was taken in the central region. Cocoon length/width ratio was calculated by following formula:

Cocoon length/width ratio = 100 × length/width

RESULTS

Performance of silkworm breeds/hybrids towards artificial parthenogenesis: Induction of parthenogenetic development in excised unfertilized eggs in bivoltine silkworm breeds has been given in Fig. 1.3. Among Japanese type bivoltine silkworm breeds, eggs of CSR_{19} expressed maximum

Table 1.2: Salient features of bisexual polyvoltine breeding lines of the silkworm, *Bombyx mori* L. with androgenetic origin

Line	Parents involved	Method adopted	Sex	Cocoon colour	Cocoon shape	Cocoon weight (g)	Cocoon shell weight (g)	Cocoon shell (%)
AGL_1	DNP_3	Androgenesis coupled with conventional breeding	Bisexual	Light-greenish yellow	Oval	1.274	0.228	17.89
AGL_2	ND_7	Androgenesis coupled with conventional breeding	Bisexual	Light-greenish yellow	Oval	1.301	0.230	17.67
AGL_3	BL_{68}	Androgenesis coupled with conventional breeding	Bisexual	Light-greenish yellow	Oval	1.336	0.230	17.22
AGL_4	NP_1	Androgenesis coupled with conventional breeding	Bisexual	Light-greenish yellow	Oval	1.347	0.237	17.59
AGL_5	DNP_5	Androgenesis coupled with conventional breeding	Bisexual	Light-greenish yellow	Oval	1.382	0.250	18.09

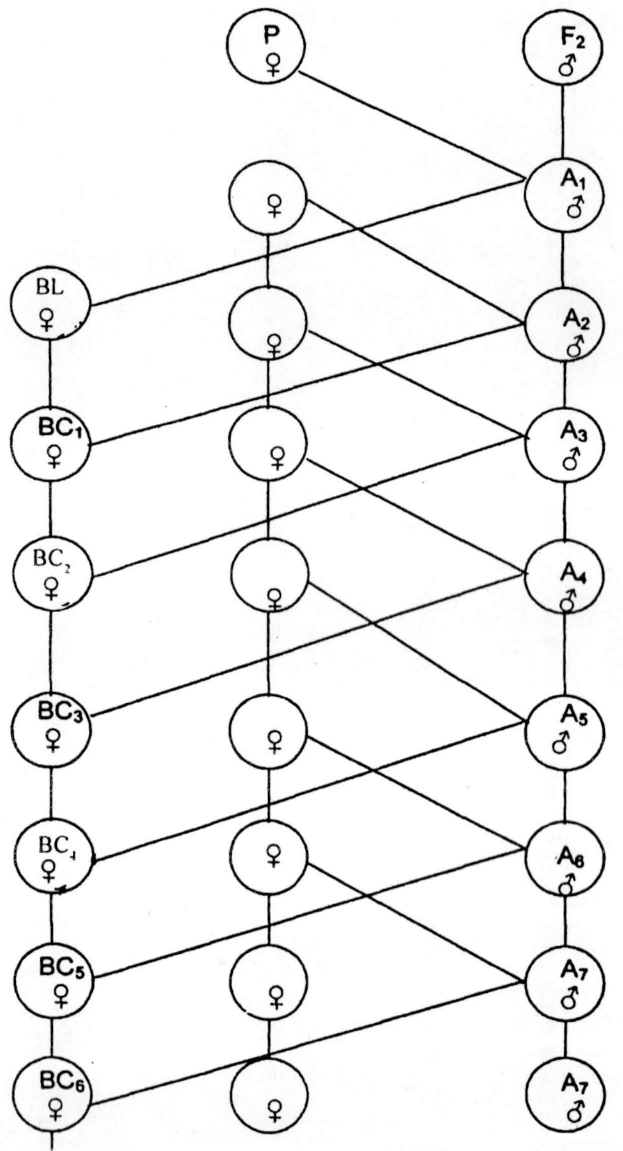

Fig. 1.2. A breeding strategy for the development of bisexual lines of the silkworm with androgenic origin;
BL - breeding line; P - Female kept as genetic marker; A - androgenetic male; BC - backcrosses.

parthenogenetic development (58.55%) followed by NB_4D_2 (51.09%) and CSR_4 (50.85%). Among Chinese type bivoltine breeds, CSR_{12} exhibited maximum parthenogenetic development (39.85%) followed by CSR_2 (39.25%) and CSR_{18} (37.17%). In general, Japanese type bivoltine breeds expressed higher response towards parthenogenetic development as compared to Chinese type.

Comparative induction of artificial parthenogenesis in excised unfertilized eggs of different bivoltine F_1 hybrids is depicted in Fig. 1.4. CSR_{19} × CSR_6 showed maximum parthenogenetic development (91.20%) followed by CSR_{19} × CSR_4 (88.56%). Hatching% was recorded maximum in CSR_{19} × CSR_4 (78.13%) followed by CSR_{19} × CSR_6 (51.05%) and CSR_6 × CSR_{26} (47.78%). Among Chinese type bivoltine hybrids, maximum parthenogenetic development was observed in CSR_{18} × CSR_{12} (81.65%) followed by CSR_3 × CSR_{17} (60.34%) and CSR_{17} × CSR_{12} (57.97%). Hatching% was maximum in CSR_{18} × CSR_{12} (10.57%) followed by CSR_2 × CSR_{27} (6.78%).

Genetic factors responsible for artificial parthenogenesis: Response towards artificial parthenogenesis greatly varied among the parental breeds and F_1 hybrids. Therefore, to know the genetic factors responsible for artificial parthenogenesis crosses were made between the breeds with the highest and lowest ability towards artificial parthenogenesis. A great deal of variation between the straight cross and its reciprocal was observed (Fig. 1.5). Maximum parthenogenetic development (51.86%) and hatching (24.48%) were observed in the cross CSR_{19} × CSR_2 (SL) possessing female component with the highest parthenogenetic ability than its reciprocal cross CSR_2 (SL) × CSR_{19} possessing female with lowest parthenogenetic ability.

Performance of silkworm hybrids towards androgenic development: Induction of androgenic development in different silkworm hybrids has been given in Fig. 1.6. Androgenic development was found more when the eggs were treated at 38 ÚC for 200 min. In bivoltine × polyvoltine hybrid, CSR_{17} × Nistari exhibited maximum androgenic development (67.85%) followed by CSR_2 (SL) × ND_7 (67.49%) and BL_{68} × CSR_{17} (66.74%). Treatment of eggs at 40 ÚC for 135 min, androgenic development was more (65.46%) in the eggs of CSR_{17} × Nistari followed by NB_4D_2 × CSR_2 (60.96%) and CSR_2 (SL) × ND_7 (59.59%).

Evaluation of silkworm breeds/hybrids through various statistical measures: In order to shortlist promising bivoltine silkworm breeds/hybrids, performance based on various statistical measures, one best parent and two better hybrids for 11 economic characters namely, fecundity, hatching, pupation rate, yield/10,000 larvae by weight, cocoon weight, cocoon shell weight, cocoon shell percentage including post cocoon parameters like filament length, reelability, raw silk percentage and neatness have been given in Table 1.3. DNB_1 exhibited its superiority for 5 characters namely, pupation rate, cocoon shell percentage, filament length, raw silk percentage and neatness when all the 3 parameters *viz.*, average performance, evaluation index value and GCA effects were taken into consideration. In addition, other 2 characters namely, hatching (based on average performance and evaluation index value) and cocoon yield (based on GCA effects) were also observed maximum in that breed. Three characters namely, cocoon yield, cocoon weight, cocoon shell weight were maximum in DNB_6 based on average performance and evaluation index value whereas fecundity in that breed was found maximum when the GCA effects for that particular character was compared with other breeds.

DNB_1 × CSR_2 was found better showing maximum values for 5 characters namely, cocoon shell weight, cocoon shell percentage, filament length, raw silk and neatness based on average performance and evaluation index value. Cocoon shell percentage and raw silk percentage were also found higher based on other two parameters *viz.*, specific combining ability and hybrid vigour. DNB_6 × CSR_2 revealed its superiority for 4 characters namely,

Table 1.3: Performance of bivoltine breeds/hybrids based on various statistical measures

	Based on average performance	Based on evaluation index values	Based on general combining ability	Based on average performance	Based on evaluation index values	Based on specific combining ability	Based on hybrid vigour	
Fecundity (no)	DNB_2 (644)	DNB_2(64.29)	DNB_8(76.33)	$DNB_8 \times NB_4D_2$(630)	$DNB_8 \times NB_4D_2$(70.21)	$DNB_4 \times NB_4D_2$(51.97)	-	$DNB_2 \times CSR_{17}$ (155.15±6.37)
Hatching (%)	DNB_1 (96.28)	DNB_1(56.91)	-	$DNB_7 \times CSR_2$ (97.32)	$DNB_7 \times CSR_2$ (65.50)	$DNB_7 \times CSR_2$ (2.65)	$DNB_7 \times CSR_2$ (36.62)	$DNB_1 \times CSR_{17}$ (147.63±6.80)
Pupation rate	DNB_1 (91.33)	DNB_1(68.79)	DNB_1 (2.60)	$DNB_6 \times CSR_2$ (98.80) $DNB_7 \times CSR_2$ (97.73)	$DNB_8 \times CSR_2$ (67.81) $DNB_7 \times CSR_2$ (65.57)	$DNB_4 \times CSR_4$ (5.68) $DNB_7 \times CSR_2$ (5.13)	$DNB_4 \times CSR_4$ (14.08)	$DNB_1 \times CSR_2$ (144.69±7.04)
Cocoon yield/10,000 larvae by wt (kg)	DNB_8 (15.63)	DNB_6 (61.43)	DNB_1 (1.19)	$DNB_6 \times CSR_2$ (17.227) $DNB_3 \times CSR_2$ (17.107)	$DNB_6 \times CSR_2$ (63.99) $DNB_3 \times CSR_2$ (62.98)	$DNB_4 \times CSR_4$ (1.46) $DNB_3 \times CSR_2$ (0.91)	$DNB_4 \times CSR_4$ (28.30) $DNB_2 \times CSR_4$ (19.81)	$DNB_8 \times CSR_2$ (170.73±7.10)
Cocoon weight (g)	DNB_6 (1.973)	DNB_6 (64.78)	DNB_4 (0.04)	$DNB_3 \times CSR_{17}$ (1.962) $DNB_7 \times CSR_{17}$ (1.913)	$DNB_3 \times CSR_{17}$(65.73) $DNB_7 \times CSR_{17}$(60.79)	$DNB_3 \times CSR_{17}$ (0.12) $DNB_8 \times CSR_2$ (0.10)	$DNB_3 \times CSR_{17}$(21.50) $DNB_4 \times CSR_4$ (16.79)	$DNB_4 \times CSR_{17}$ (161.01±7.61)
Cocoon shell weight (g)	DNB_6 (0.380)	DNB_6 (63.01)	DNB_1(0.024)	$DNB_1 \times CSR_2$ (0.390) $DNB_3 \times CSR_{17}$ (0.383)	$DNB_1 \times CSR_2$ (63.93) $DNB_3 \times CSR_{17}$(60.97)	$DNB_3 \times CSR_{17}$(0.029) $DNB_8 \times CSR_2$ (0.025)	$DNB_4 \times CSR_4$ (21.46) $DNB_3 \times CSR_{17}$(18.94)	$DNB_7 \times CSR_{17}$ (161.10±7.65)
Cocoon shell %	DNB_1 (21.22)	DNB_1 (64.84)	DNB_1 (0.94)	$DNB_1 \times CSR_2$ (21.54) $DNB_1 \times CSR_4$ (21.00)	$DNB_1 \times CSR_2$ (78.33) $DNB_1 \times CSR_4$ (58.39)	$DNB_2 \times NB_4D_2$ (0.71) $DNB_1 \times CSR_2$ (0.65)	$DNB_7 \times CSR_4$ (5.81) $DNB_1 \times CSR_2$ (5.73) $DNB_1 \times CSR_4$ (5.00)	$DNB_1 \times CSR_4$ (167.98±7.66)
Filament length (m)	DNB_1 (870)	DNB_1 (63.62)	DNB_1(44.54)	$DNB_1 \times CSR_2$ (977) $DNB_8 \times CSR_2$ (963)	$DNB_1 \times CSR_2$ (65.91) $DNB_6 \times CSR_2$ (63.67)	$DNB_2 \times CSR_4$ (89.21) $DNB_7 \times NB_4D_2$(79.32)	$DNB_7 \times NB_4D_2$(23.50) $DNB_2 \times CSR_4$ (20.63)	$DNB_3 \times CSR_2$ (165.67±7.89)
Reelability (%)	DNB_7 (83.5)	DNB_7 (60.64)	DNB_4 (2.35)	$DNB_7 \times CSR_2$ (86.1) $DNB_8 \times CSR_2$ (84.9)	$DNB_7 \times CSR_2$ (68.24) $DNB_8 \times CSR_2$ (65.38)	$DNB_4 \times NB_4D_2$ (3.89) $DNB_2 \times CSR_{17}$ (3.66)	$DNB_4 \times CSR_2$ (8.79) $DNB_3 \times CSR_2$ (8.44)	$DNB_4 \times CSR_2$ (159.14±7.90)
Raw silk %	DNB_1 (15.70)	DNB_1 (65.03)	DN3₁ (1.14)	$DNB_1 \times CSR_2$(17.10) $DNB_1 \times CSR_4$(17.05)	$DNB_1 \times CSR_2$(74.25) $DNB_1 \times CSR_4$(73.68)	$DNB_1 \times CSR_4$(0.69) $DNB_1 \times CSR_2$(0.67)	$DNB_1 \times CSR_4$(12.61) $DNB_1 \times CSR_2$(9.71)	$DNB_4 \times CSR_4$ (195.76±7.91)
Neatness	DNB_1 (92)	DNB_1 (59.52)	DNB_1 (0.68)	$DNB_1 \times CSR_4$ (93) $DNB_1 \times CSR_2$ (92)	$DNB_1 \times CSR_4$ (65.91) $DNB_1 \times CSR_2$ (60.39)	$DNB_3 \times NB_4D_2$ (2.51) $DNB_4 \times CSR_{17}$ (2.21)	$DNB_4 \times CSR_{17}$ (3.94) $DNB_3 \times NB_4D_2$ (2.03)	$DNB_7 \times CSR_2$ (163.36±7.96)

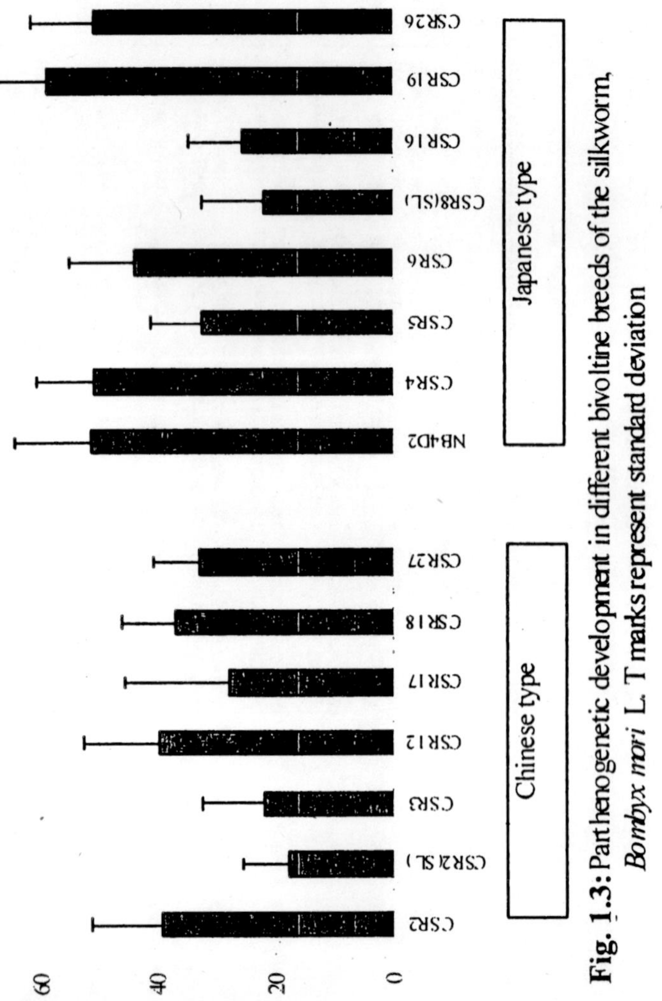

Fig. 1.3: Parthenogenetic development in different bivoltine breeds of the silkworm, *Bombyx mori* L. T marks represent standard deviation

pupation rate, cocoon yield, filament length and reelability based on average performance and evaluation index value. Though, the hybrid did not rank for hybrid vigour among the two better heterotic hybrids, comparatively higher SCA effects were found for other two characters namely, cocoon weight and cocoon shell weight. $DNB_7 \times CSR_2$ exhibited maximum values for 3 characters namely, hatching, pupation rate and reelability based on average performance and evaluation index value where higher SCA effects were observed for hatching and pupation rate. Hatching was also found highly significant when hybrid vigour analyzed. $DNB_1 \times CSR_4$ exhibited higher performance for 3 characters namely, cocoon shell percentage, raw silk and neatness based on average performance and evaluation index value. Hybrids were also evaluated for cocoon size uniformity to know their impact from commercial view point. Hybrid cocoons with standard deviation less than 8 were considered uniform in cocoon size and listed in the table in ascending orders.

Laboratory performance of short-listed bivoltine hybrids: Four short-listed bivoltine hybrids namely, $DNB_1 \times CSR_2$, $DNB_1 \times CSR_4$, $DNB_6 \times CSR_2$, and $DNB_7 \times CSR_2$ along with two additional hybrids *viz.*, $DNB_6 \times DNB_1$ and $DNB_7 \times DNB_1$ were further evaluated in the laboratory. Performance of newly developed hybrids was statistically compared with $CSR_2 \times CSR_4$ taken as control to select promising hybrid combination. Comparative performance of newly developed hybrids along with $CSR_2 \times CSR_4$ has been presented in Table 1.4. Among 6 bivoltine hybrids, 2 hybrids namely, $DNB_1 \times CSR_4$ and $DNB_7 \times DNB_1$ showed higher average evaluation index values of 55 and 58 respectively. Highly significant ($p < 0.01$) increase for 5 characters *viz.*, fecundity (586), cocoon shell weight (0.387 g), filament length (1050 m), denier (2.66 d) and raw silk percentage (17.57%) and significant ($p < 0.05$) increase for 2 characters *viz.*, cocoon shell percentage ((21.50%) and neatness (91.7 p) were recorded in the hybrid $DNB_1 \times CSR_4$. The hybrid $DNB_7 \times DNB_1$ recorded highly significant ($p < 0.01$) increase for 8 characters *viz.*, cocoon yield/

Table 1.4: Performance of the short listed bivoltine hybrids in the laboratory

Hybrids	Fecundity (no)	Hatching (%)	Total larval span (hrs)	Pupation (%)	Yield/10000 larvae by wt. (kg)	Cocoon wt. (g)	Cocoon shell wt. (cg)	Cocoon shell (%)	Filament length (m)	Denier (d)	Reel-ability (%)	Raw silk (%)	Neatness (p)	Avg. E. I. Value†	Cocoon size uniformity ††
$DNB_1 \times CSR_2$	583**	97.20	528	96.00	16.29	1.749	0.370	21.18	967	2.80	80.6	16.70**	91	42	146.41±6.06
$DNB_1 \times CSR_4$	586**	98.13*	510	96.67	17.25	1.798	0.387**	21.50*	1050**	2.66**	82.1	17.57**	92*	55	165.96±6.64
$DNB_6 \times DNB_1$	534	97.69	510	97.60	17.07	1.775	0.378	21.27	1012**	2.75	82.0	17.63**	91	50	166.81±6.85
$DNB_6 \times CSR_2$	557**	98.568	510	95.07	16.96	1.787	0.386**	21.62**	993	2.85	80.7	17.47**	88	47	169.79±7.01
$DNB_7 \times DNB_1$	535	97.07	510	97.07	17.23	1.809	0.394**	21.78**	1015**	2.80	82.0	17.60**	92**	53	165.48±7.07
$DNB_7 \times CSR_2$	536	97.14	510	96.40	17.76**	1.880**	0.411**	21.85**	1077**	2.65**	82.0	17.77**	92**	58	163.56±6.92
$CSR_2 \times CSR_4$	535	97.57	510	96.40	17.17	1.765	0.369	20.93	991	2.75	82.0	16.60	91	45	168.04±7.08
CD@5%	**15**	**0.88**	–	**1.29**	**0.26**	**0.060**	**0.010**	**0.47**	**11**	**0.03**	**0.9**	**0.39**	**0.9**	.	.
CD@1%	**21**	**1.23**	–	**1.80**	**0.36**	**0.084**	**0.014**	**0.66**	**15.29**	**0.04**	**1.31**	**0.55**	**1.23**	.	.
SE ±	**4.82**	**0.29**	–	**0.42**	**0.08**	**0.02**	**0.00**	**0.15**	**3.55**	**0.09**	**0.30**	**0.13**	**0.29**	.	.

† Avg. E. I. value indicate the average evaluation index values for positive characters; †† Cocoon size uniformity indicates mean of cocoon length / width ratio ± SD of 100 cocoons.

* and ** denote significant difference at 5% and 1% level respectively.

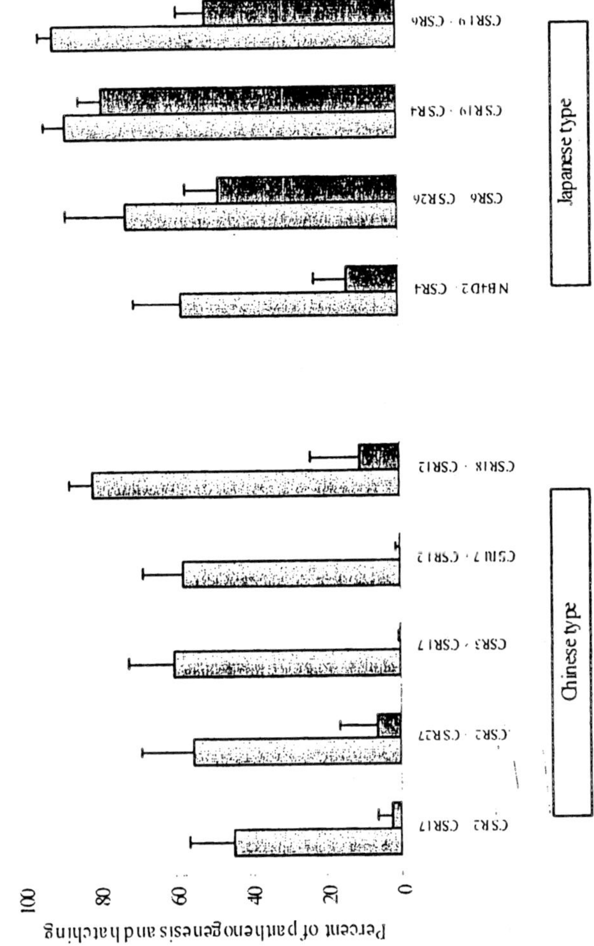

Fig. 1.4: Parthenogenetic development in bivoltine hybrids of the silkworm, *Bombyx mori* L. T marks represent Standard Deviation.

10,000 larvae by weight (17.760 kg), cocoon weight (1.880 g), cocoon shell weight (0.411 g), cocoon shell percentage (21.85%), filament length (1077 m), denier (2.65 d), raw silk percentage (17.77%) and neatness (92 p). Two hybrids namely, $DNB_6 \times CSR_2$ and $DNB_7 \times CSR_2$ exhibited significant increase for at least 5 out of 13 characters studied. Fecundity (583) and raw silk percentage (16.70%) were found highly significant in the hybrid $DNB_1 \times CSR_2$. $DNB_6 \times DNB_1$ showed significant increase for filament length (1012 m) and raw silk percentage (17.63%). All the hybrids exhibited uniformity in cocoon size with intermediate in shape between dumbbell and oval possessing standard deviation less than 8.

Preliminary field trial of promising bivoltine hybrids: Two selected bivoltine hybrids *viz.*, $DNB_1 \times CSR_4$ and $DNB_7 \times CSR_2$ along with control $CSR_2 \times CSR_4$ were further evaluated both in the laboratory and with a few farmers located in Karnataka. Rearing results of 400 dfls each of the selected bivoltine hybrids tested with the farmers recorded an average cocoon yield of 65.450, 69.770 and 60.350 kg/100 dfls, cocoon weight of 1.789, 1.827 and 1.741 g, cocoon shell weight of 0.371, 0.399 and 0.354 g and cocoon shell percentage of 20.74, 21.85 and 20.35% in $DNB_1 \times CSR_4$, $DNB_7 \times CSR_2$ and $CSR_2 \times CSR_4$ respectively. Results showed superiority of $DNB_7 \times CSR_2$ over other hybrids both in the laboratory as well as in the field (Table 1.5).

DISCUSSION

Development of silkworm breeds by conventional breeding has played a vital role in upgrading both quality and quantity of silk produced (Datta, 1984). Since most of the quantitative characters in silkworm are governed by polygenes, their inheritance shows variation and, therefore, more emphasis is being paid for selection of sikworm breeds based on their phenotypic expression (Nagaraju, 1998). Sometimes, a suitable phenotype may not exhibit a suitable genotype and due to low heritability, the subsequent progeny may lose its unique genotype. Artificial parthenogenesis

Table 1.5: Performance of the promising bivoltine hybrids in the the field

Hybrids	No. of dfls	No. of farmers	Yield/100 dfls (kg)	Cocoon wt. (g)	Cocoon shell wt. (cg)	Cocoon shell (%)	Cocoon rate / kg (Rs)
$DNB_1 \times CSR_4$	400	4	65.45 (8.45)	1.789 (2.76)	0.371 (4.80)	20.74 (1.92)	165=00 (2.17)
$DNB_7 \times CSR_2$	400	4	69.77 (15.61)	1.827 (4.94)	0.399 (12.71)	21.85 (7.37)	175=00 (8.36)
$CSR_2 \times CSR_4$ (Control)	400	4	60.35	1.741	0.354	20.35	161=50

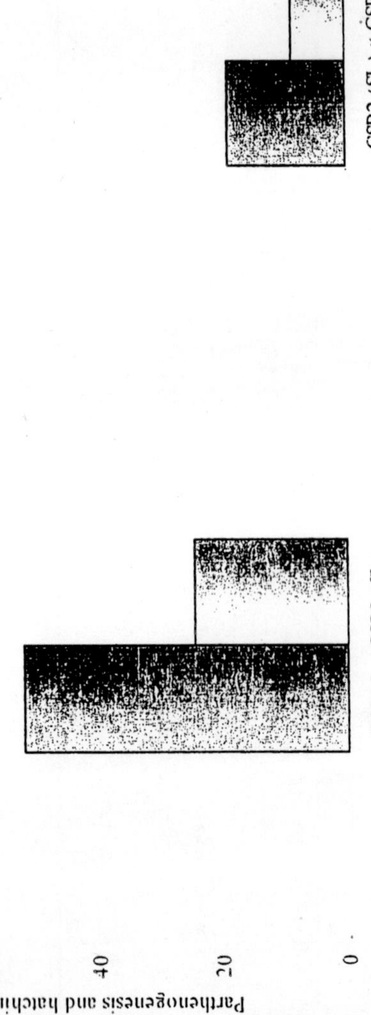

Fig. 1.5: Parthenogenetic development in straight and reciprocal crosses of different hybrids of the silkworm, *Bombyx mori* L.

enables one to produce from one outstanding individual hundreds of parthenoclones each of which is an exact genotypical copy of its parent (Astaurov, 1957; Strunnikov, 1975). In sexual reproduction, the offsprings receive only a random half of alleles from each parent and the results are not predictable accurately (Seidel and Brackett, 1981). Application of new breeding strategies like parthenogenesis and androgenesis would be beneficial to the silk industry in the development and cloning of homozygous silkworm breeds with either entirely females (completely heterozygous) via ameiotic parthenogenesis or predominantly males (homozygous) via androgenesis to improve the selection efficiency (Strunnikov, 1983, 1986; Retnakaran and Percy, 1985; Takei *et al.*, 1990). Though, the practical significance of artificial parthenogenesis and androgenesis has been realized, less attention has been given to explore the possibility of utilizing these strategies for the development of superior silkworm breeds/hybrids found in India. Of late, attempts are being made at Central Sericultural Research and Training Institute, Mysore to develop superior breeds/hybrids of the silkworm through the application of artificial parthenogenesis and androgenesis (Singh *et al.*, 2004; Gangopadhyay and Singh, 2007).

Response of different silkworm breeds towards parthenogenetic development has been studied. Among the different bivoltine and polyvoltine breeds, Japanese type bivoltine silkworm breeds showed pronounced parthenogenetic development (Gangopadhyay and Singh, 2006). An increased tendency towards parthenogenesis was observed in hybrids obtained from a mother moth having a high tendency of parthenogenesis. Breeds with higher parthenogenetic ability were crossed to establish parthenogenetic character in the lines. Three bivoltine breeds *viz.*, DNB_1, DNB_6 and DNB_7 were developed. The bisexual line DNB_1 was characterized by sex-limited characteristics with white oval cocoons while DNB_6 and DNB_7 were characterized by entirely female parthenoclones possessing white dumbbell cocoons. Hybrids were prepared by crossing the developed silkworm breeds

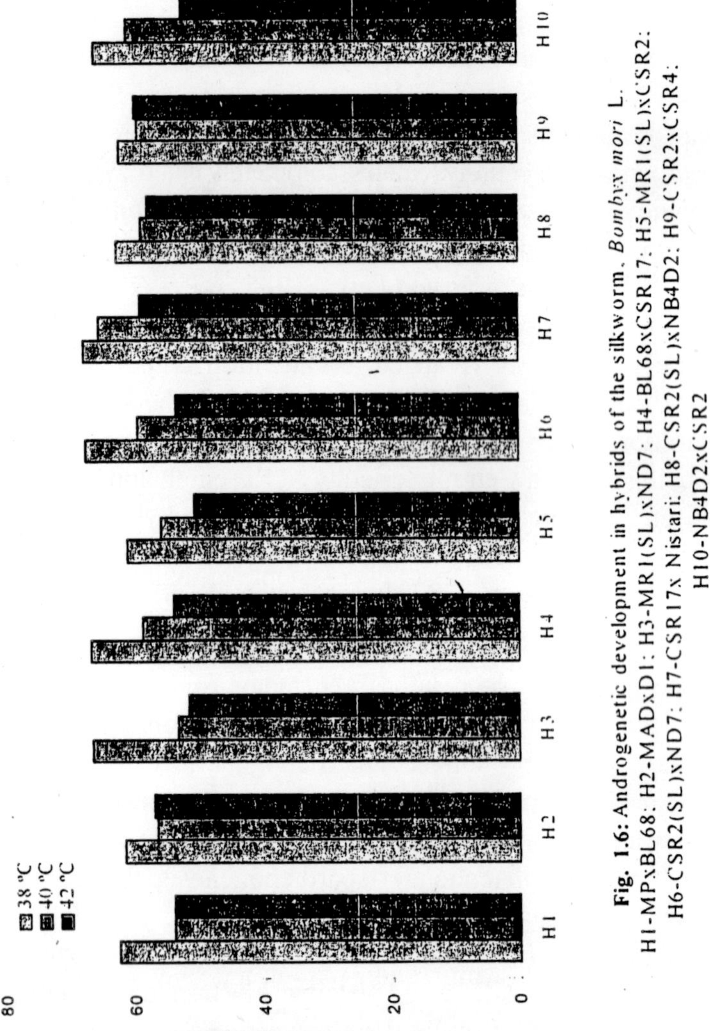

Fig. 1.6: Androgenetic development in hybrids of the silkworm, *Bombyx mori* L.
H1-MPxBL68; H2-MADxD1; H3-MR1(SL)xND7; H4-BL68xCSR17; H5-MR1(SL)xCSR2; H6-CSR2(SL)xND7; H7-CSR17x Nistari; H8-CSR2(SL)xNB4D2; H9-CSR2xCSR4; H10-NB4D2xCSR2

and productive bivoltine breeds. Evaluation of the developed breeds/hybrids was carried out through various statistical measures like analysis of combining ability, hybrid vigour and cocoon size uniformity. Studies showed more combining ability, hybrid vigour and cocoon size uniformity in the new hybrids.

Induction of androgenesis was performed in different silkworms to select potential breeds. Polyvoltine hybrids showed higher androgenic development as compared to bivoltine hybrids. Nistari, a polyvoltine breed possessing dominant gene for golden yellow cocoon colour with marked larvae was utilized as genetic marker to identify the androgenic male individual. Females of Nistari were crossed with the males of the hybrid $BL_{68} \times BL_{69}$ possessing plain larvae and oval shaped greenish yellow cocoons. Induction of androgenesis was performed by exposing the oviposited eggs at hot air of 38 °C for 200 min. Both marked and plain larvae were observed. Plain larvae were identified as androgenic individuals. Sex of the plain larvae at pupal stage was further determined. All the pupae derived from plain larvae were exclusively males. Backcrossing was adopted utilizing androgenic males to introgress homozygosity in the breeding lines. By utilizing dispermic androgenesis, bisexual silkworm lines have been isolated (Xu *et al.*, 1997). Nacheva *et al.*, (1999) have developed some bisexual lines of the mulberry silkworm, *B. mori* with androgenetic origin. Further studies of the breeds developed via parthenogenesis and androgenesis will be carried out through DNA fingerprinting to know the level of homozygosity.

Application of parthenogenesis and androgenesis in the mulberry silkworm has several advantages like development of homozygous lines within short period, individuals with less phenotypic variability and hybrids with high viability, combining ability and hybrid vigour. The newly developed silkworm breeds may further be utilized for the development of superior breeds by the silkworm breeders.

REFERENCES

Astaurov, B. L. (1940). Artificial parthenogenesis in the silkworm (*Bombyx mori* L.) (Experimental study). *Academia delle Scienze Dell'urss*, pp. 221–240.

Astaurov, B.L. (1957). High temperature as a tool for controlling development and sex determination: A review of studies in artificial parthenogenesis, androgenesis and elimination of embryonic diapause in the silkworm, *Bombyx mori* L. *Proc. Zool. Soc.* Calcutta, Mookerjee Memoir, 1: 29-56.

Datta, R.K. (1984). Improvement of silkworm races (*Bombyx mori* L.) in India. *Sericologia*, 24 (3): 393-415.

Dznealaidze, A.N. & Tabliashvili, T.S. I. (1990). Cloned hybrid lines of the silkworm. *Shelk.* 4: 7-8.

Gangopadhyay & Singh, R. (2006). Parthenogenetic development in excised eggs of F_1 hybrids of the silkworm, *Bombyx mori* L. *Sericologia*, 46(2): 229-232.

Gangopadhyay & Singh, R. (2007). Development and evaluation of parthenoclones of the mulberry silkworm, *Bombxy mori* L. *Sericologia*, In: Press.

Kempthorne, O. (1957). *An Introduction to Genetics Statistics*. John Willy and sons, Inc. New York, pp. 208-314.

Mano, Y., Kumar, N.S., Basavaraja, H.K., Reddy, N.M. & Datta, R.K. (1993). A new method to select promising silkworm breeds/combinations. *Indian Silk*, 31: 53.

Nacheva, Y., Petkov, N., Tzenov P., Malinova, K. & Gensu Su, C. (1999). Breeding of bisexual lines of the silkworm, *Bombyx mori* L. with androgenetic origin. *Bull. Indian Acad. Seric.*, 3(1): 36-41.

Nagaraju, J. (1998). Silk yield attributes – correlations and complexities. *Silkworm Breeding*. Sreerama Reddy, G. (ed.), Oxford and IBH Publishing Co. Pvt. Ltd. New Delhi. pp. 168-185.

Nagoya, H., Okamoto, H., Nakayama, I., Araki, K. & Onozato, H. (1996). Production of androgenetic diploids in amago salmon, *Onchorhynchus masou*. *Fisheries Sci.*, (Tokyo) 62: 380-383.

Plugaru, I.G., Plugaru, R. I., Golouko, V. A., Stotskii, M.I., Spiridonova, T. I., Klimenko V.V. (1993). Perspective hybrids of silkworm, *Bombyx mori* L. found during parthenogenesis. *Bull. Acad. Repub. Moldova Biol. Sci.*, 3: 12-16.

Singh, R., Gangopadhyay, D.D.R., Choudhary, N., Kariappa, B.K. & Dandin, S.B. (2004). Development of a polyvoltine breed – BL_{67} (Pg) of the silkworm, *Bombyx mori* L. with parthenogenetic origin. *Int. J. Indust. Entomol.*, 9(1): 41-46.

Retnakaran, A. & Percy, J. (1985). Fertilization and special modes of reproduction. In: *Comprehensive Insect Physiology, Biochemistry and Pharmacology*. Kerkut, G..A. and Gilbert, L.I. (eds.), Vol. 1, Pergamon Press, New York. pp. 242-253.

Seidel, G..E. Jr. & Brackett, B.G.. (1981). Perspectives on Animal Breeding. In: *New Technologies in Animal Breeding*. Brackett, B.G., Seidel, G..E. Jr. and Seidel, S.M. (eds.), Academic Press, London. pp.3-9.

Strunnikov, V. A. (1975). Sex control in silkworms. *Nature*, 255: 111–113.

Strunnikov, V. A. (1983). *Control of silkworm reproduction, development and sex.* Mir Publishers, Moscow. pp. 1-280.

Strunnikov, V.A. (1986). Nature of heterosis and combining ability in the silkworm. *Theor. Appl. Genet.*, 72 (5): 503–512.

Strunnikov. V.A. (1995). *Control over reproduction, sex and heterosis of the silkworm.* Harwood Academic publishers, Moscow, pp. 1-343.

Takei, R., Nakagaki, M., Kodoira, R. & Nagashima, E. (1990). Factor analysis on the parthenogenetic development of ovarian eggs in the silkworm, *Bombyx mori* L. (Lepidoptera: Bombycidae). *Appl. Entomol. Zool.*, 25 (1): 43-48.

Tazima, Y. & Onuma, A. (1967). Experimental induction of androgenesis, gynogenesis and polyploidy in *Bombyx mori* L. by treatment with CO_2 gas. *J. Seric. Sci. Japan*, 36: 286-292.

Tazima, Y. (1964). *The Genetics of the Silkworm*. Logos Press (Academic Press), London, pp. 1-253.

Whiting, R. (1955). Androgenesis as evidence for the nature of x-ray induced injury. *Radiation Res.*, 2: 71-78.

Xu, A.N., Li, M.W., Fang, A., Fei, M.H. & Huang, T.T. (1997). Isolation of a self breed line of *Bombyx mori* L. by means of dispermic androgenesis. *Sericologia*, 37: 199-204.

Ye, Y., Jiang, W.Q. & Chen, R. (1989). Studies on cytology of crosses between nucleus and cytoplasm in distant hybridization of fishes. *Acta-Hydrobiologica Sinica*, 13: 234-239.

BREEDING OF NEW THERMO-TOLERANT BIVOLTINE SILKWORM HYBRID ATR16 X ATR29 AND ITS PERFORMANCE AT FARMERS' LEVEL

Abad A. Siddiqui, M.A. Khan, T.P.S. Chauhan, R. Lochan** and Mir Nisar Ahmad*

ABSTRACT

Though during recent years a number of productive bivoltine breeds/hybrids were developed at different institutes in India for both sub-tropical and temperate regions but these hybrids are suitable for rearing during favourable seasons and require optimum temperature and humid conditions for the expression of their full potential. Due to the very causes, there are several reports of moderate to heavy silkworm crop losses during hotter months. Without the availability of hardy bivoltine silkworm races, it is also difficult to introduce one or two additional sericulture crops in summer and monsoon seasons particularly in the state of Uttar Pradesh. To breed new bivoltine silkworm genotypes with hardiness against high temperature and humidity as prime objective and with moderate productive trait, a breeding project was initiated in 1998 at RSRS, Sahaspur, Dehradun. Twenty bivoltine and eight Biv. X Multi crosses were served as breeding materials to isolate new breeds/hybrids and the breeding methodologies advocated by Chinese breeders were followed. During F1 to F3 generation, mass method of breeding was adopted to increase the variation in breeding populations and to get the more gene recombinants, this has facilitated in selection of desirable batches and

Key words: Thermotolerant, Bivoltine, Silkworm, Breeding Hybrids.

Central Sericultural Research and Training Institute, Pampore-181101, J&K, India.

* RSRS, Miran Sahib, Jammu, J&K.

** P2 Basic Seed Farm, Dehradun-248001.

E-mails: siddiquibad@yahoo.co.in; csrtipper@vsnl.com; mirnisarahmad@gmail.com individuals in later generations. From F4 to F10 generation, cellular rearing with pedigree/line selection method was adopted. Subsequently, three Chinese type lines viz. ATR6, ATR16 and ATR28 and three Japanese type viz. ATR11, ATR13 and ATR29 have been evolved. Hybrids of these thermo-tolerant breeds have the pupation rate around 90% with shell ratio between 21-23% which is about 25% more than control ruling hybrid. Among the six new hybrids, ATR16 X ATR29 has excelled over the other and subjected to hybrid testing during 2003,2004, 2005 and 2006 in Dehradun, H.P. and in U.P. An average cocoon yield in ATR16 X ATR29 was 32 kg/100 dfls in summer season whereas in control the yield was only 12 kg .In the present paper, the methods of evolution of new hardy breeds and its performance at farmers level have been discussed in detail.

Key words: Silkworm, breeding, hybrids, thermo-tolerance.

INTRODUCTION

For competing in international market and to sustain the sericulture industry, India should produce more of quality bivoltine cocoons with gradable silk. However, production of bivoltine cocoons could not get the expected momentum because still there is instability in bivoltine crops while the rearing of multivoltine crops is easier as multitvoltine x bivoltine combinations are sturdy towards high temperature and high humidity. Due to this, still in India major portion of cocoons produced is from multivoltine x bivoltine combinations. Therefore, it is very essential to develop bivoltine silkworm breeds suitable for rearing under high temperature and high humid conditions.

Most of the bivoltine breeds developed in India till recently though are highly productive and spin 2A and 3A grade silk but require optimum inputs of quality mulberry leaf and good rearing conditions to exhibit full potential. These races and their hybrid when reared in sub-optimal conditions suffer a heavy loss due to different diseases and succumb to them. So far, harsh weather conditions were not given due attention by the breeders and thermo-tolerant breeds could not be developed for rearing during unfavorable seasons. Datta, (2000) has the opinion that the hot climatic conditions of tropical and sub-tropics, particularly the summer and monsoon seasons are not suitable for rearing spring specific races. It was also suggested by different experts of sericulture that for Indian sericulture industry development of races for sub-optimal conditions is of immense importance. Therefore, a breeding programme was under taken at RSRS, Sahaspur, Dehradun to develope thermo-tolerant bivoltine silkworm breeds suitable for rearing during summer and monsoon seasons of north-western states of India. Ahsan and Rao (1997) are also of the opinion that it is the need of the time for increasing the pace of vertical growth of sericulture and to develop high yielding sturdy breeds. With the availability of robust breeds, farmers of hotter areas will be able to take up the sericulture crops in summer and monsoon also.

MATERIALS AND METHODS

Exotic and indigenous bivoltine races available at germplasm bank of RSRS, Sahaspur were evaluated against high temperature and high humid conditions to select comparatively tolerant races to these conditions as parent materials. 44 bivoltine races were reared in summer season in 1997 and the races which have recorded the pupation rate above 70% were selected. Selected races were crossed to raise single crosses and as such 10 bivoltine crosses of Oval X Oval parents viz. CC1 x NB7, CC1 x DUN7, CC1 x SF19,CC1 x KA, NB7 x Dun 7,NB 7 x SF19,NB7 x KA, Dun 7 x SF19 and DUN7 x KA and SF19 x KA and 10 crosses of Constricted x Constricted parents viz Taichoan x Shinkuryaku, Taichoan x JD6,Taichoan X NB4D2,Taichaon x Su, Shinkuryaku x JD6, Shinkuryaku x NB4D2, Shinkuryaku x Su, JD6 x NB4D2,JD6 x Su and NB4D2 x Su were prepared and utilized as breeding resource materials. In additions to these single crosses eight bivotine x multivoltine (White cocoon and diapusing) crosses namely Pam111 X SF19, NB18X ADG6, Shinkurytaku x ADG6,Taichoan x ADG6,Pam111 x TW, Dun17 x TW, Taichoan x TW and Su x TW were also raised and utilised as breeding population to isolate new lines.

Mass rearing method was adopted during F1 to F3 (Year 1998-1999) generations for obtaining more gene recombinants and to create variations in breeding populations so that desirable individuals which have hardiness(tolerance) as well as moderate cocoon characters could be selected in later generations of breeding. Cellular rearing was resorted to between F4 to F9 (Year 2000 and 2001) generation and pedigree/line separation method of breeding was adopted. Each breeding line during these mid-generations was reared in 5 cellular batches during hotter months of summer and monsoon. Selection was made both at batch and individual level and the batches with pupation rate above 70% and shorter larval period were only selected for furthering the breeding generations. In selected batches, desirable individuals(cocoons) were selected through single cocoon assessment. Selection in each generation was made in such a way that the balance between negatively correlated characters such as pupation rate and shell weight. and shell ratio remain maintained. Since the pupation rate is the indication of healthiness of a breed, selection was mainly based on pupation rate of the lines. However, other economic characters viz.. fecundity, hatching %, larval period, yield/10000 larvae, single cocoon weight, single shell weight and shell ratio and post cocoon traits viz. filament length, denier, reelability, neatness were also considered during selection. The lines which have not recorded any improvement in pupation rate and not responded to selection were culled during breeding and finally six lines have been fixed at F10 generation (Summer 2001).Selection was also made on larval marking, cocoon shape and size. Hybrid testing of new breeds was conducted in summer and monsoon season of 2002 as per the evaluation method suggested by Mano et.al.,(1993) and ranked as per manos evaluation Index. In the breeding populations which are obtained from the crosses of bivoltine x multivoltine parents, in segregating generations cocoons which were towards bivoltine characteristics and were selected for the next breeding generation to maintain the productivity character of bivoltine intact. Most promising hybrid ATR16 x ATR29 was tested along with control during summer and autumn in Himanchal Pradesh and in summer in Dehradun, in the years 2003, 2004,2005 and 2006.

RESULTS AND DISCUSSION

The out come of the present breeding work is six thermo-tolerant silkworm bivoltine breeds and among them three breeds viz. ATR6, ATR16 and ATR28 are Chinese type (plain larvae and oval cocoons) and ATR11, ATR13 & ATR29 are Japanese type (marked larvae,constricted cocoons). Improvement in pupation rate over the ruling bivoltine races was achieved. Status of different economic characters of new breeds is summarised in Table 2.1. Perusal of data revealed that pupation rate in new thermo-tolerant breeds ranged between 77.2% (ATR11) and 86.0%(ATR29) and the shell weight ranged from 0.317 gm to 0.365 g with shell ratio between 18.11% (ATR29) and 20.37% (ATR6) in harsh climatic conditions. Filament length in these six breeds ranged between 900 m to 1060 m with denier between 2.60 and 3.00.

For hybrid testing of new breeds, three Chinese type breeds were crossed with three Japanese type breeds because the heterosis in this type of crosses (Chinese x Japanese) is high due to the difference in their genetic constituent, and for commercial crop also, this sort of combinations are utilized. The average performance of nine new hybrids under rearing room temperature between 29 to 33 ^{0}C along with control SH6 x NB4D2 is summarized in table 2. In new hybrids pupation rate ranged between 75.6% and 90.2% while in control the pupation rate was only 72%. In shell weight also all new hybrids showed improvement and excelled over control and it was between 0.265 g to 0.395 g and shell ratio was recorded from 17.56% to 23.19% against 0.23g and 16.39% respectively recorded in control. Improvement in post cocoon parameters was also, achieved and filament length ranged from 940 M to 1053 M. Ranking of nine hybrids for different economic characters which was done by Manos evaluation index revealed that hybrid ATR16 x ATR29 recorded the highest average index value of 60.29% followed by ATR 6 x ATR29 . The average index value in control hybrid was only 33.88% (Table 3) The index value of pupation rate which indicates the survival ability was also highest in ATR16 x ATR29 (Index value 66.0%). In productivity also two hybrids viz ATR16 x ATR29 and ATR6 x ATR29 were excelled over the others.

Keeping in view the over all performance, uniformity in larval marking, cocoon shape, moulting behaviour ATR16 x ATR29 has been recommended for commercial use in summer and monsoon seasons up to rearing room temperatures of 33-34 °C (out side temperature around 36-37 °C).

Chinese method of breeding for developing thermo-tolerant bivoltine breeds was adopted. As per Chinese breeders, it is very important for breeders to set up suitable breeding environment according to the breeding goal during fixation of new lines. The environmental conditions during evolution of new lines should be similar to those conditions in which the resulted breeds would be exploited later on at farmers level. To breed the healthy genotypes, environmental conditions should be such that full expression of character of healthiness against high temperature could be obtained so that selection of those lines that have potential to survive better under harsh conditions can be done. In the present study, lines were reared during filial generation during summer and monsoon seasons. This helped in natural culling of weak individuals/families and facilitated the selection of tolerant lines. This also helped by infusing tolerance in new lines. For the present breeding work, two type of crosses i.e. bi x bi and ii) x bi x moltivoltine were utilized to evolve hardy breeds. Bi x multivoltine crosses were utilized with an aim to introduce the character of healthiness of the multivoltine in productive bivoltine races. Different Chinese breeders (Heyi *et. al.*,1991) viewed that by the gene recombination of two types of parents and successive selection in filial generations good characters of both parents can be unified in new genotype which possess high resistance to adversity of bad environment and moderate silk out put. From bi x bi crosses ATR6, ATR11, ATR13 and ATR16 were developed whereas ATR28 and ATR29 were developed from Bi

Table 2.1: Status of different characters of new thermo-tolerant breeds in harsh seasons.

Traits	*ATR6*	*ATR11*	*ATR13*	*ATR16*	*ATR28*	*ATR29*
Larval marking	Plain	Marked	Marked	Plain	Plain	Marked
Cocoon shape	Oval	Constricted	Constricted	Oval	Oval	Constricted
Fecundity	536	540	480	542	542	536
Larval period (D:H)	23:06	22:12	23:00	23:06	23:00	23:06
Yield/10000 larvae(kg)	12.85	12.09	12.66	13.10	13.31	13.80
Pupation rate	81.3	77.2	78.8	84.5	80.0	86.0
Single cocoon wt.(g)	1.792	1.735	1.674	1.720	1.692	1.750
Single shell wt.(g)	0.365	0.345	0.319	0.343	0.326	0.317
Shell ratio (%)	20.37	19.88	19.06	19.94	19.27	18.11
Avg. filament length(mts)	1030	1000	960	900	1060	930
Denier	2.85	2.60	3.00	2.90	2.80	2.60
Reelability(%)	83.0	89.0	80.0	83.0	80.0	86.0

Table 2.2 : Status of different characters in new hybrids

Traits	ATR6	ATR11	ATR13	ATR16	ATR28	ATR29
Larval marking	Plain	Marked	Marked	Plain	Plain	Marked
Cocoon shape	Oval	Constricted	Constricted	Oval	Oval	Constricted
Fecundity	536	540	480	542	542	536
Larval period (D:H)	23:06	22:12	23:00	23:06	23:00	23:06
Yield/10000 larvae(kg)	12.85	12.09	12.66	13.10	13.31	13.80
Pupation rate	81.3	77.2	78.8	84.5	80.0	86.0
Single cocoon wt.(g)	1.792	1.735	1.674	1.720	1.692	1.750
Single shell wt.(g)	0.365	0.345	0.319	0.343	0.326	0.317
Shell ratio (%)	20.37	19.88	19.06	19.94	19.27	18.11
Avg.Filament Length(mts)	1030	1000	960	900	1060	930
Denier	2.85	2.60	3.00	2.90	2.80	2.60
Reelability (%)	83.0	89.0	80.0	83.0	80.0	86.0

Table 2.3: Evaluation Index values of different hybrids for different characters and their ranking

Hybrids	Fecun dity	Yield/ 10,000 larvae (Kg)	Pupation Rate (%)	Single cocoon wt.(g)	Single cocoon wt.(g)	Shell ratio (%)
ATR6XATR11	533	13.42	86.4	1.756	0.373	21.24
ATR6X ATR13	500	12.7	80.3	1.642	0.306	18.63
ATR6X ATR29	498	14.11	85.4	1.720	0.366	21.28
ATR16 XATR11	516	13.02	83.2	1.703	0.395	23.19
ATR16X ATR13	509	13.79	86.1	1.629	0.342	20.99
ATR16X ATR29	537	14	90.2	1.724	0.36	20.88
ATR28XATR11	484	12.06	80.5	1.507	0.292	19.38
ATR28XATR13	500	11.99	78.4	1.429	0.273	19.1
ATR28XATR29	475	11.46	75.6	1.509	0.265	17.56
SH6 XNB4D2 (Control)	503	11.9	72	1.403	0.23	16.39

x Multi crosses. Performance of a breed for different characters depends on gene and environment interaction which later plays a vital role for the success of the breeds/hybrids at commercial level. Therefore, it was suggested that the breeding environment should be set up as per breeding target. In this present study because it was the main objective that new hybrids will be utilized in harsh season, therefore, rearing of breeding lines throughout all filial generations was conducted in summer and monsoon seasons. It is a genetic phenomenon that survival is negatively correlated with silk content and it is difficult to breed new breed with high survival rate as well as high silk out put. Thus, for summer and monsoon seasons, it is possible to develope bivoltine breeds with better pupation rate and moderate silk content. Japanese scientists observed that resistance to high temperature is a heritable character and the silkworm lines selected at high temperature/humidity perform better than the lines selected under normal conditions. They opined that it is possible to breed silkworm lines tolerant to high temperature(Koundinya *et al.*, 2006).Chinese breeders have long back evolved bivoltine breeds for their tropical areas (Sohn *et. al.*,1987 and Shao, 1989). In Thailand also, a number of robust bivoltine breeds such as 4792 and BL were developed (Lethi Kim, 1987).CSR&TI, Mysore has also developed thermo-tolerant breeds CSR18 and CSR19 by exposing breeding materials to high temperature using sericatron rearing chamber. Begum *et al.* (1999) have also successfully bred two hardy bivoltine strains A3 x 935E and A3 x 916B with high survival and moderate silk recovery.

Results of field trails have been summarized in table 4 show that the average cocoon yield /100 dfls in ATR16 x ATR29 was 32.2 kg while in control the yield was only 17.0 kg . Introduction of new thermo-tolerant hybrid ATR16 x ATR29 will be helpful in increasing the number of bivoltine crops in summer seasons in the state of U.P., Uttarakhand, Himanchal Pradesh and J&K and will boost the return of the farmers form sericulture.

Table 2.4: Results of field trials of ATR16 X ATR 29 along with control

	Evaluation Index values of different Characters					**Avg. Evaluation Index**
Hybrid	Yield /10000 larvae	Pupation rate	SCW	SSW	SR%	
ATR6XATR11	56.29	58.75	62.46	60.17	57.16	58.97
ATR6X ATR13	48.41	47.12	53.23	47.26	43.58	47.92
ATR6X ATR29	63.83	56.84	59.55	58.82	57.36	59.28
ATR16 XATR11	51.91	52.65	58.17	64.41	67.30	58.89
ATR16X ATR13	60.33	58.18	52.17	54.20	55.86	56.15
ATR16X ATR29	62.63	66.00	59.87	57.67	55.28	60.29
ATR28XATR11	41.42	47.50	42.28	44.57	47.48	44.65
ATR28XATR13	40.65	43.50	35.96	33.91	46.03	35.41
ATR28XATR29	34.86	38.16	42.45	39.37	38.02	38.57
SH6 X NB4D2 (Control)	39.67	31.30	33.86	32.62	31.93	33.88

REFERENCES

Ahsan, M.M & Rao, R.P. (1997). Progress through silkworm breeding in India. Base paper, *Current technology seminar,* 18-19 Sept., CSR&TI, Mysore.

Begum, A.N., Ahsan, M.M. & Datta, R.K. (1999). Breeding of two bivoltine A3 X 935 E and A3 X 916B of silkworm *Bombyx mori* L. for high survival and moderate silk out productivity, Korean. *J. Seric. Sci.,* 41(2): 94-101.

Datta, R.K. (2000). Silkworm breeding in India: Present status and new challenges. *National conference on strategies for sericulture research and development*, 18 Nov., CSR&TI, Mysore.

Heyi Sina, Y..H., Jiang, D.A. & Dai, P. (1991). Breeding of the silkworm varieties for summer and autumn rearing Xinhua, Qiuxing and their hybrids. *Canye Kexue*, 17: 200-207.

Koundinya, P.R., Somasundram, P., Kumersan, P., Sinha, R.K. & Rajeurs, S. (2006). *India Silk*, 45(5): 7-8.

Kim, L. (1987). Evolution of bivoltine strain of silkworm *Bombyx mori* L. for summer rearing in Vietnam. *Sericologia*, 27(3): 437-441.

Ruli, S. (1989). Breeding of Lantin x Baiyan, a summer-autumn silkworm variety. *Sci. Seric.*, 15: 125-129.

Sohn, K.W., Hong, K.W., Ryu, K.S., Chai, S.R., Manyi Kim, K.Y., Lee, S.P. & Kuan, Y.H. (1987). Breeding of Dalsongjkam a sex limited larval marking and high silk yielding silkworm variety for summer-autumn rearing. *Res. Rep. Rural Dev. Adm.* (Suweon). p. 29.

Mano, Y., Kumar, N.S., Basavaraja, H.K., Malareddi, N. & Datta, R. K. (1993). A new method to select promising silkworm breeds combinations. *India Silk*, 31: 53.

3

MULBERRY SHOOT FEEDING TO 5TH STAGE SILKWORMS PROVES TO BE MUCH MORE LUCRATIVE THAN INDIVIDUAL MULBERRY LEAF FEEDING FOR HIGHER COCOON PRODUCTIVITY, WITH SPECIAL REFERENCE TO AUTUMN SEASON IN SUB-TROPICS OF JAMMU REGION (J&K)

M.A. Khan, S.K. Raina, T.P.S. Chauhan, S.L.Dhar and B.B. Bindroo

ABSTRACT

Normaly, the traditional silkworm rearing in northern states of India is carried out by feeding the silkworms with individually picked up mulberry leaves, which demands intensive labour involvement, particularly for day to day leaf plucking, 3-4 times/day silkworm feeding in trays, daily silkworm bed cleaning etc. The 7-8 days' rearing of 5th stage silkworms alone costs nearly 50% of the total cost of labour required for silkworm rearing.

Further, the fast withering of individually plucked leaves during their transportation, preservation and even in rearing trays leads to loss of quality of leaf and ultimately diminishing the palatability of leaf to silkworms, causing them to under feed which results in poor and a feeble cocoon crop. Since, the freshness of leaves with shoots lasts for a longer duration and the shoots retain more moisture and supply the same to attached leaves until the leaves are fully consumed by the silkworms, a comparative study between mulberry shoot feeding and the individually picked up leaf feeding was taken up during both spring and autumn season.

Regional Sericultural Research Station, Miransahib, Jammu-181101;
E-mails:csrtipper@vsnl.com

During spring season. the cocoon yield with mulberry shoot was recorded to be 5.03% more over the yield recorded with individual leaf feeding. whereas in autumn season, a gain of 19.7% yield (with mulberry shoots) was recorded. Hence, the mulberry shoot feeding to late age silkworms should be advocated and popularized for its adoption in the field for commercial rearing.

Key words: Mulberry. Cocoon productivity, Shoot feeding, Seasons, Commercial rearing

INTRODUCTION

Normally, the silkworm rearing in India is carried out by feeding the silkworms with individually picked up mulberry leaves demanding intensive labour involvement, particularly for daily leaf plucking /picking, 3-4 times silkworm feedings, daily silkworm bed cleaning etc. upto the 5th. stage of silkworms.

Such activities are not only tedious and tiring for silkworm rearers, but also time consuming and more expensive in terms of lablour for them.

The 7-8 days' rearing of 5th stage silkworms alone costs about 50% of the total cost of labour required during the whole silkworm rearing i.e, from brushing of silkworms until the spining stage. Any shortage of labour during the 5th stage of silkworms leads to poor and a feeble cocoon crop or even to total failure of the harvest.

Low cocoon productivity is caused by a number of factors, of which the quality of leaf fed to silkworms plays a major role. The fast withering/wilting of individually picked up leaves during their transportation from mulberry field/garden to the rearing place, their preservation and even in the silkworm rearing beds leads to loss of quality of leaf and ultimately diminish the palatability of leaf to silkworms causing them to underfeed, which results in a poor cocoon crop.

Mostly, as observed during the surveys undertaken, the sericultural farmers in Jammu region feed the 5th stage silkworms only 2-3 times/day instead of established four feedings a day, as they experience the picking of individual mulberry leaves in large quantum, and thereafter feeding the voraciously eating 5th stage larvae to be tedious, laborious and time consuming jobs.

Since, the mulberry shoots keep supplying the moisture/water contents to the mulberry leaves attached to them, maintaining their freshness for a longer duration until the leaves are fully consumed by the silkworms, a comparative study between mulberry shoot feeding and the individually picked up mulberry leaf feeding was taken up during both spring and autumn seasons to assess the impact of mulberry shoot feeding of 5th stage larvae on cocoon productivity in comparison to individual mulberry leaf feeding.

MATERIAL AND METHODS

To asses the impact of feeding 5th stage silkworms with mulberry shoots on cocoon productivity in comparison to individual mulberry leaf feeding, the 5th stage silkworms were fed with mulberry shoots and individual mulberry leaves side by side simultaneously during both spring and autumn season in 3 replications each with 300 larvac of SH6 x NB4D2 hybrid.

Frequency of feeding was maintained @ four feedings a day. Silkworm bed cleaning under the mulberry shoot feeding treatment was restricted to only once during the entire 5th stage larvae i.e., 2nd day of 5th stage by using two ropes of strings.

RESULTS AND DISCUSSIONS

Cocoon crop results were found to be better in case of mulberry shoot feeding of 5th stage silkworms than the individual mulberry leaf feeding of 5th stage silkworms, as the cocoon yield was recorded to be more both in spring as well as autumn season.

Data recorded during spring season (Table 3.1) reveals that the cocoon yield in case of mulberry shoot feeding was more, being 18.100 kg/10,000 larvae against 17.233 kg/10,000 larvae i.e. a gain of 5.03% yield (with mulberry shoots) over the yield obtained with individual mulberry leaf feeding.

During autumn season (Table 3.2), the cocoon yield was recorded to be 17,160 kg/10,000 larvae in case of mulberry shoot

Table 3.1: Comparative rearing performance of silkworms fed with individual mulberry leaf and mulberry shoots. (Hybrid: SH6 x NB4D2) Season: Spring

Sl.No.	Parameters	Feeding with individual mulberry leaf	Feeding with mulberry shoots
1.	Average yield /10,000 larvae		
	By No.	9766	9867
	By wt.(Kg.)	17.223	18.100
2.	Average single cocoon wt. (g)	1.816	1.853
3.	Average single shell wt. (g)	0.316	0.345
4.	Average shell ratio %	17.400	18.618
Yield gain (%) in weight with shoot feeding over individual leaf feeding		5.03%	

Table 3.2: Comparative rearing performance of silkworms fed with individual mulberry leaf and mulberry shoots. (Hybrid: SH6 x NB4D2) Season: Autumn

Sl.No.	Parameters	Feeding with individual mulberry leaf	Feeding with mulberry shoots
1.	Average yield /10,000 larvae		
	By No.	9500	9900
	By wt.(Kg.)	14.330	17.160
2.	Average single cocoon wt. (g)	1.613	1.784
3.	Average single shell wt. (g)	0.283	0.347
4.	Average shell ratio %	17.540	19.45
Yield gain (%) in weight with shoot feeding over individual leaf feeding		19.70%	

feeding, whereas an yield of 14.330 kg/10,000 larvae was recorded with individual mulberry leaf feeding. A much more yield gain of 19.70% was recorded by adopting mulberry shoot feeding of 5th stage silkworms over the yield gained under the individual mulberry leaf feeding treatment.

CONCLUSION

On the basis of the results obtained from the study, it can be concluded that the cocoon production can easily be raised by nearly 20%, particularly in autumn season, if the 5th stage silkworms are fed with mulberry shoots instead of individual mulberry leaves. Hence, the mulberry shoot feeding to 5th stage silkworms should be advocated and popularized for its adoption in the field for commercial rearing.

REFERENCES & SUGGESTED READINGS

Dandin, S.B., Jayaswal, J. & Giridhar, K. (2000). *Hand book of Sericulture Technologies*. Central Silk Board, Bangalore.

Datta, R.K., Basavaraja, H.K & Mano, (1996). *Manual on Bivoltine Rearing, Race Maintenance and Multiplication*. CSR&TI, Mysore.

Khan, R.A., Raina, S.K., Koul, S., Bindroo, B.B., Misri, S.S. & Quadri, S.M.H. (2003). An appropriate way of mulberry shoot preservation. *Indian Silk* (July), pp. 13-14.

Sekharappa, B.M., Gururaja, C.S., Raghuraman, R. & Dandin, S.B. (1991). *Shoot feeding for late age silkworms*. Karnataka state Sericultural, Research and Development institute.

Siddappa, Ji C., Muraleedhara, M.R. & Chikkavenkateshappa (1992). Advantages of rearing silkworms throughout on mulberry shoots. In: *Nat. Conf. Mulberry Seric., Res.* CSR&TI, Mysore – Dec. 10 -11, p. 64.

SUSTAINABILITY IN QUALITY COCOON PRODUCTION IN MANDYA DISTRICT OF KARNATAKA – THROUGH ADOPTION OF JICA TECHNOLOGIES

*G.B. Singh, C.K. Kamble, M. Venkateswara Rao and T.S. Tsuchiya**

ABSTRACT

After the success of first and second phase of JICA project (Japan International Co-operation Agency) Government of India with the assistance of Government of Japan launched another project called PEBS (Project for strengthening of Extension System of Bivoltine Sericulture) . The project was mainly confined to three of Southern states viz; Karnataka, Andhra Pradesh and Tamilnadu. Initially surveys were conducted to select the Technical Service Center (TSC) and farmers of the TSCs. In Karnataka, eight TSCs were selected and one young age silkworm rearing center popularly known as Chawki rearing center (CRC) was established with each TSC. Mandya district of Karnataka was having three JICA selected TSCs i.e. Mandya, Koppa and Thoresettyhally. The study was conducted in three TSCs of Mandya. Initial survey was conducted with 10 farmers in each TSC as an indicator farmers and average cocoon yield was maximum (57.9 kg) in Koppa TSC and minimum was 52 kg in Mandya TSC. Similarly, highest and lowest cocoon melting percentage was 16.8 and 11.6 in Mandya and Koppa TSCs. After introduction of JICA programme, the worms were reared up to 2nd age in CRC and distributed among selected farmers. Crops were supervised at all crucial stages such as brushing, 4th and 5th age and during mounting and harvesting. A total of 34 batches rearing were conducted. During the year 8, 9, 12 and 5 batches were reared under 1st, 2nd, 3rd and 4th crop respectively. During the course of rearing 116, 118,143 and 48 farmers were selected to

Central Sericultural Research and Training Institute, Sreerampura, Mysore-570008, INDIA.
E-mail: sgbcsrti@rediffmail.com
*JICA Export.

conduct the rearing in four crops. 27400, 24525, 30475 and 11100 DFLs were reared in 1st, 2nd, 3rd and 4th crop respectively. Hatching percentage ranged from 89 to 90%. During 1st crop actual yield and yield per 100 DFLs was 18626 and 68.3 kg respectively. 16661 kg actual yield and 67.6 kg average yield was recorded for 2nd crop. For 3rd and 4th crops actual yield were 20717 and 6701 kg and average yield was 68.2 and 60.3 kg respectively. Percent of melted cocoons ranged from 4.6 to 5.6% Rate of cocoon/kg was in between Rs. 130 to 160. Data clearly indicated that average cocoon production increased 25.6, 16.5 and 18.7 per cent in Mandya, Koppa and T.S. Hally TSCs. This increase in cocoon production is attributed due to drastic reduction (236, 133 and 175 per cent in Mandya, Koppa and T. S. Hally, TSCs respectively) of melting cocoons and increase in cocoon weight (8 to 12.5% increase).

The present study indicated that field impact and adoption of new bivoltine sericulture technologies leads to production of high quality of cocoons. It is concluded that the sustainability in quality cocoon production is due to adoptability of all improved technologies, constant supervision of the crops and creating awareness among farmers.

INTRODUCTION

India is the second largest silk producing, silk consumer and also the world's largest importer of silk. Almost 95% of raw silk production is either from traditional multivoltine or multivoltine and bivoltine hybrids. Presently, the estimated demand of improved raw silk (bivoltine) in the country is about 6000 MT. To meet this demand and to improve the quality of raw silk produced to that of the international grade, it is inevitable for us to go for large scale production of bivoltine raw silk. Though, the production of bivoltine silk was introduced as early as 1970, it could not get wide acceptance due to various reasons. India being a tropical country, bivoltine sericulture could not get popularity despite a quantum jump in silk production and quality of silk remains inferior (Datta, 1984). Considering the magnitude of expansion of mulberry area, it is apparent that cocoon production has not recorded proportional growth and there is a wide gap between the laboratory and actual yield obtained by the farmers. In case of bivoltine, only 51% potential was realized in the field (Angadi, 2002).

By sensing these problems, Government of India launched a massive project i.e. Bivoltine Sericulture Technology Development (BSTD) during 1992-1997. The project was assisted by the Government of Japan under the Japan International Co operation Agency (JICA) The technologies for production of bivoltine silk were developed and tested in field in a limited scale. Immediately after completion of BSTD project, second phase of the project started from 1997 to 2002 and called as Popularizing Practical Bivoltine Sericulture Technology (PPPBST). In this project, all the newly developed technologies were tested in all three southern states i.e. Karnataka, Andhra Pradesh and Tamilnadu. After the grand success of 2nd phase of JICA, both Indian and Japanese Government decided to start 3rd phase of JICA project and this project was aimed to strengthening the bivoltine extension system in India. Under this project, studies were concentrated on the extension system in sericulture and its impact on silk growth. Project was also centered towards strengthening of extension system.

MATERIALS AND METHODS

The lists of the Technical Service Centers (TSC) were given by the respective Department of Sericulture. Japanese experts along with scientists of Central Sericultural Research and Training Institute, Mysore visited the TSCs for final selection. After selecting the TSCs next step was to identify the young age silkworm rearing center popularly known as Chawki Rearing Centers (CRCs). One TSC was having at least one CRC. In Karnataka 8 TSCs and 8 CRCs were identified to start the project. In Mandya district of Karnataka, 3 TSCs namely Mandya, Thoresettyhally and Koppa were selected. After the selection of TSC and CRC, scientist surveyed the farmers of respective TSC and selection was made. Criteria for selecting the farmers were listed below:

Farmer should have

1. Minimum one acre mulberry garden
2. V1 or S36 irrigated mulberry garden

Table 4.1: TSC wise crop performance before implementation of JICA programme.

Name of TSC	No of farmers	No of DFLs	Hatching %	Actual yield (Kg)	Yield / 100 Dfls (kg)	Melting %	Rate / kg of Cocoons (Rs.)	Cocoon weight (g)	Shell weight (g)	Shell (%)
Mandya	10	1200	90	628	52.3	16.8	127.0	1.698	0.375	22.09
Koppa	10	1800	89	1035	57.5	11.6	129.0	1.765	0.379	21.48
T.S. Hally	10	1250	87	690	55.2	14.6	120.0	1.700	0.368	21.65
Total/Avg	30	4250	88	2353	55.37	14.3	125.0	1.721	0.374	21.74

Table 4.2: TSC wise crop performance after implementation of JICA programme

Name of theTSC	No of Batch	No of farmers	No of DFLs	Hatc-hing %	Actual yield (Kg)	Yield / 100 Dfls (kg)	Melting %	Rate / kg of Cocoons (Rs.)	Cocoon weight (g)	Shell weight (g)	Shell (%)
Mandya	2	21	5500	90	3657	66.5	5.8	147.0	1.773	0.378	21.33
	2	23	4200	90	2702	64.3	5.2	140.0	1.600	0.345	21.56
	6	56	11600	90	7698	66.4	5.0	147.0	1.754	0.394	22.44
	2	19	3500	92	2296	65.6	4.0	155.0	1.735	0.395	22.74
Total/Avg	**12**	**119**	**24800**	**90.5**	**16353**	**65.7**	**5.0**	**147.2**	**1.715**	**0.378**	**22.01**
Koppa	4	67	14500	92	9668	66.7	5.5	136.0	1.918	0.426	22.21
	4	60	12000	91	7976	66.5	4.9	128.0	1.960	0.433	22.09
	4	53	10875	90	7396	68.0	4.5	159.0	1.950	0.445	22.82
	2	18	3950	91	2646	67.0	5.0	159.0	2.040	0.430	21.08
Total/Avg	**14**	**198**	**41325**	**91**	**27686**	**67.0**	**5.0**	**145.5**	**1.967**	**0.434**	**22.05**
T.S. Hally	2	28	7400	87	5301	71.6	5.6	139.0	1.794	0.375	20.90
	3	35	8325	91	5983	71.9	5.7	123.0	1.744	0.409	23.46
	2	34	8000	91	5623	70.3	5.1	152.0	1.805	0.406	22.49
	1	11	3650	85	1759	48.2	4.8	168.0	1.757	0.396	22.54
Total/Avg	**8**	**108**	**27375**	**88.5**	**18666**	**65.5**	**5.3**	**145.5**	**1.775**	**0.397**	**22.35**

3. Wider spacing of mulberry garden
4. Willing to receive chawki reared worms
5. Separate silkworm rearing house
6. Shoot rearing technology
7. Minimum one year bivoltine silkworm rearing experience
8. Willing to accept the technical advices provided by the experts
9. Separate mounting space

Initial survey was conducted to study the performance of the selected farmers. During this process, if farmers had few lacunae, they were advised to improve it immediately. After 35 days of initial survey, a final survey was conducted to select the farmers. After the selection of farmers young age silkworm rearing center was established. Rearing capacity of the farmer was decided based on the availability of mulberry garden and rearing space. Accordingly, laying indents were collected and eggs were transported on the day of brushing from grainage to CRC. Laying were brushed at CRC under the supervision of experts. Only CSR hybrids were utilized for the rearing. Chawki, late age and mounting and harvesting of cocoons were done as per Kawakami and Yanagawa (2003). Close monitoring was conducted during young age rearing. After the certification (By the team of experts), chawki worms were distributed among farmers during 2nd moult. Crops were supervised at all the crucial stages by the team of the experts. Cocoons were harvested on 6th/7th day and marketed on next day after sorting and deflossing the cocoons. In addition to the crop supervision, group discussion, exhibition, field day and other mass communication methods were also adopted to create awareness among the farmers.

RESULTS

Initial data of the rearing performance of three TSCs were collected and presented in Table 4.1. Ten farmers data from each TSC are presented. 1200, 1800 and 1250 DFLs data were collected from Mandya, Koppa and T.S. Hally TSCs respectively. Hatching ranged between 87 to 90%. Average cocoon yield was maximum (57.5 kg) in Koppa TSC and minimum (52.3 kg) was in Mandya TSC and 57.5 kg yield was recoreded from T.S. Hally TSC. Melting percentage of cocoons was maximum (16.8%) in

Mandya and it was minimum (11.6%) in Koppa TSC. Cocoon weight of all three TSC ranged between 1.698 g to 1.765 g, Cocoon shell weight between 0.368 g to 0.379 g and Cocoon shell percentage between 21.48% to 22.09%.

CROP PERFORMANCE

TSC wise and batch wise crop performance is given in Table 4.2. Summary of the crop performance and improvement over initial performance is presented in Table 4.3. A total of 34 batches were reared in the year. Maximum 14 batches reared in Koppa TSC and minimum 8 in T.S. Hally TSC. Total 425 farmers were included in present study. Maximum 198 farmers of Koppa and minimum 108 farmers of T.S. Hally were included. Number of DFLs reared was 24,800; 41,325 and 27,375 from Mandya, Koppa and T.S. Hally TSC respectively. Total 93,500 DFLs were reared in 34 batches and 62,705 kg cocoons harvested. Yield/100 DFLs was 65.7 kg, 67.0 kg and 65.5 kg from Mandya, Koppa and T.S. Hally TSCs respectively. It was evident that significant improvement in cocoon yield and reduction in occurrence of melting cocoons observed in all three TSCs. When this cocoon yield was compared with initial yield, data reveals that maximum improvement of 26.53% was recorded from Mandya TSC followed by T.S. Hally TSC (18.66%) and minimum improvement of 16.53% was recorded from Koppa TSC. The average cocoon melting per cent reduces from 14.3% to 5.1%. TSC wise reduction in melting is 16.8% to 5%; 11.6% to 5.0% and 14.6 to 5.3% in Mandya, Koppa and T.S. Hally TSC respectively. Cocoon fetch higher rate (Rs. 20 –25 /kg) compared to initial cocoon rate.

COCOON ASSESSMENT

Cocoons were assessed on the day of harvesting. The result of cocoon assessment is presented in Table 4.4. The average of four rearing cocoon assessment of Mandya TSC shows that cocoon weight was 1.715 g and cocoon shell weight 0.378 and cocoon shell percentage is 22.02. Cocoon weight of Koppa TSC was 1.967 g and cocoon shell weight was 0.434 g and cocoon shell percent was 22.05. For T.S. hally TSC it was 1.775 g, 0.397 g and 22.35%. No significant difference was found in cocoon characters.

Table 4.3: Summary of crop performance and improvement after implementation of JICA programme

Name of the TSC	No of Batch	No of farmers	No of DFLs	Hatching %	Actual yield (Kg)	Yield / 100 Dfls (kg)	% of improve-ment in cocoon yield	Melting %	Reduction in melting (%)	Rate / kg of cocoons (Rs.)
Mandya	12	119	24800	90.5	16353	65.7	26.53	5.0	236	147.25
Koppa	14	198	41325	91.0	27686	67.0	16.53	5.0	132	145.50
T, S. hally	8	108	27375	88.5	18666	65.5	18.66	5.3	175	145.50
Total/Avg	**34**	**425**	**93500**	**90.0**	**62705**	**66.1**	**20.19**	**5.1**	**181**	**146.1**

Table 4.4: Cocoon assessment and reeling performance- TSC wise

Name of the TSC	Cocoon weight (g)	Cocoon shell weight (g)	Cocoon shell percentage	Reelability (%)	Filament length (m)	Renditta (good cocoons)	Neatness points
Mandya	1.715	0.378	22.02	78	950	7.58	90
Koppa	1.967	0.434	22.05	82	1100	6.90	95
T.S. Hally	1.775	0.396	22.54	74	935	7.25	91
Average	**1.819**	**0.402**	**22.20**	**75**	**995**	**7.24**	**92**

REELING PARAMETERS

Sample of cocoons were sent for reeling test at Central Silk Technological and Research Institute, Bangalore. Data are presented in Table 4.4. Average of four crops, reeling test showed that average filament length of Mandya, Koppa and T.S. Hally TSCs were 950, 1100 and 935 m respectively. Reelability was 78, 82 and 74% from the batches of cocoons harvested from Mandya, Koppa and T.S. Hally TSCs respectively. Renditta was 7.58, 6.90 and 7.25 for three TSCs in same order. Highest neatness point (95) was recorded from the Koppa TSC cocoons.

DISCUSSION

From the above results it is obvious that all three TSC achieved significant progress after the implementation of JICA programme. Therefore proper implementation and planning of any programme bring a significant achievement in production of quality cocoons in enough quantity. The adoption of technologies brings the potential and positive changes in cocoon production (Rahmatulla and Geethadevi, 2000; Kumareshan *et al.,* 2002).

Visibly or invisibly, the new silkworm breed CSR had brought lot of changes with the farmers both in terms of social as well as economic benefits. The maximum improvement in average cocoon yield was recorded in Mandya TSC and minimum in Koppa TSC because before implementation of JICA scheme Koppa was having maximum cocoon yield and T.S. Hally was minimum. It was mentioned in Seri Business, (2003) that there is a potential limit for improvement and maximum improvement can be made in the case where production is minimum and vice versa with very fewer exceptions.

Reason for higher cocoon yield in Koppa is due to more cocoon weight. Contrary to this, higher cocoon melting percentage of Mandya & T. S. Hally TSC leads to low cocoon yield (Kawakami and Yanagawa, 2003). It is evident that higher cocoon weight and low melting percentage lead to increase in quality cocoon production. Lakshmanan and Geethadevi (2005) observed that the net profit earned from bivoltine cocoon production is much higher than cross-breed rearing. However, the major concern is frequent changes in cocoon price, which is not economical to quality cocoon producers. This calls for suitable price policy measures to protect bivoltine rearers, which would discourage import of silk from China in the long run. Therefore, all possible measures should be taken to produce quality cocoons.

Regular contacts with sericulture extension officials and officers, experts leads to significant improvement in quality cocoon production. In addition, participation in extension programmes such as group discussion, film shows, demonstrations, exhibitions and field days had enabled the farmers to gain knowledge about new technologies for adopting the same.

Mass media participation is very important for transfer of any technology and its adoption but it has not yet established its existence for a noticeable level (Geetha *et al.*, 2005).

The personal and socio economic characters of the farmers such as caste, social participation and level of knowledge in sericulture were found to have a significant association with the constraints faced by them in sericulture (Dhane and Dhane, 2004). However, these problems are more conspicuous in Mandya TSC and least in Koppa TSC this is another reason for good and comparatively low crop results in both TSCs.

REFERENCES

Angadi, B.S. (2002). *Workshop on promotion of bivoltine sericulture technology* under PPPBST project. Central Sericultural Research and Training Institute, Mysore. March 7-8. pp. 50-58

Datta, R. K. (1984). Improvement of silkworm races (*Bombyx mori* L.) in India. *Sericologia*, 24(3): 393-415.

Dhane, V.P. & Dhane, A.V. (2004). Constraints faced by the farmers in mulberry cultivation and Silkworm rearing. *Indian J. Seric.*, 43 (2): 155-159.

Geetha, G.S., Srinivasa, G. & Prakash, N.B.V. (2005). Comparative performance of CSR hybrids and traditional cross breed- A study in Mysore and Mandya districts of Karnataka. *Indian J. Seric.*, 44(1): 13-17.

Kawakami, K. & Yanagawa (2003). *Illustrated hand book on silkworm disease control technology.* JICA, Central Sericultural Research and Training Institute, Mysore, India.

Kawakami, K. & Yanagawa (2003). *Illustrated working process of new bivoltine silkworm rearing technology.* JICA, Central Sericultural Research and Training Institute, Mysore, India.

Kumereshan, P., Bhogesha, K., Tsuchiya, H. Vijayaprakash, N.B. & Kawakami, K. (2002). Advances in Indian Sericulture Research. *Proceedings of the National Conference on strategies for Sericulture Research and Development*, 16-18 November 2000, pp. 500-504.

Lakshmanan, S. & Geethadevi, R.G.. (2005). A comparative analysis of economics of bivoltine and cross-breed cocoon production in Mandya district of Karnataka- A micro level evidence. *Indian J. Seric.*, 44(2): 179-182.

Rahmatulla, K. & Geethadevi, R.G.. (2000). Sericulture: An apt. venture for rural betterment.

Seri Business Manual (2003). Published by Central silk Board. Edited By Member Secretary Vol. I & II. pp. 326-248.

STUDIES ON COCOON SIZE VARIABLES VIS-A-VIS REELING TRAITS OF BIVOLTINE SILKWORM (*BOMBYX MORI* L.) GERMPLASM

B. Mohan, N. Balachandran, M. Muthulaksmi, R.K. Sinha and S.M.H. Qadri

ABSTRACT

Central Sericultural Germplasm Resources Centre (CSGRC), Hosur, India has an assemblage of 432 *Bombyx mori* silkworm germplasm accessions, which are being characterised, evaluated, conserved and utilized. These silkworm germplasm exhibit wide variability both for qualitative and quantitative traits. Qualitative traits more often have either positive or negative correlation, which needs to be studied from the data base point of view. Study on the cocoon size variables and reeling traits was carried out with 101 bivoltine silkworm germplasm based on single cocoon reeling data with the objectives (i) to analyze the degree of correlation of cocoon length, width and L/W ratio on total filament length, non-broken filament length, (ii) to understand whether the shape of cocoons influences the reeling process, (iii) to find the cocoon shape uniformity among the accessions and (iv) to group the best accessions based on cocoon shape uniformity and longest filament length with less breakage during reeling cycle. The results revealed that 13 accessions (BBE-175, BBE-178, BBE-179, BBE-180, BBE-181, BBE-182, BBE-185, BBE-186, BBE-187, BBE-192, BBE-197, BBI-205 and BBE-270) showed more uniformity for cocoon shape, as its standard deviation and co-efficient of variation for cocoon length, width and L/W ratio were less than the overall mean values 1.886, 1.90 and 21.727: and 5.620, 9.990 and 12.360 respectively. These accessions

Central Sericultural Germplasm Resources Centre, Hosur-635 109, Tamil Nadu, India.
E-mails: bawcsgrc@yahoo.co.in; lakshamcsgrc@yahoo.co.in.

also showed an average total filament length more than 1000 meters and less breakage during reeling cycles. indicating the superiority in filament strength. Correlation analysis showed that cocoon length, width and L/W ratio do not have statistically significant positive or negative correlation with total filament length or non-broken filament length. These 13 silkworm germplasm have elongated and oval cocoon shape and this study suggests that cocoon shape (oval or elongated) should be given more importance in breeding plan than considering the cocoon size (cocoon length. width).

Key words: Cocoon size. bivoltine. Green plasm, Reeling traits.

INTRODUCTION

Studies made on the variability in quantitative traits of mulberry silkworm *Bombyx mori* L. (Chatterjee *et al.*, 1993; Lie, 1996; Rao *et al.*, 1997) indicate that the shape of the cocoon influences silk reelability (Mano, 1994 and Nakada, 1994). Variations in cocoon shape of parental silkworm breeds have been studied by Hiraishi (1912); Nakada (1989, 1991, 1992a and b; 1993); Nakada *et al.*, (1991); Katsuki and Nagasawa (1917). The relationship of cocoon shape to the linear measurement of larval body is also reported. Cocoon spinning process also has direct relationship with size and shape of cocoons that influence the reeling process (Miura *et. al.*,1995, 1997). Cocoon shape uniformity influences the reeling process (Mano, 1994) and the cocoon shape uniformity has direct relation with reeling process using semi-automatic and automatic reeling machines. In the present study, evaluation of cocoon shape uniformity using different cocoon shape variables such as cocoon length, cocoon width and cocoon length-width ratio and correlation of these cocoon shape variables with total reelable filament length and non-broken filament length were carried out with 101 bivoltine silkworm germplasm. Based on the results obtained, the degree of correlation between cocoon shape variables with total filament length and breakage were analysed and the germplasm accessions with more uniformity in cocoon shape, high filament length and less breakage were grouped and listed.

MATERIALS AND METHODS

A total of 101 bivoltine silkworm germplasm accessions, which are being currently maintained at CSGRC, Hosur were taken for the present study. Cocoon shape was categorized as oval, elongated with faint constriction and elongated with deep constriction. Cocoon shape variables viz., length and width of cocoon were recorded using electronic vernier calipers considering a sample of 50 cocoons/accession. Based on the cocoon length and width data, length-width ratio were worked out for individual accessions. Reeling parameters such as total filament length and numbers of breaks during reeling process were recorded using automatic epprovette for each accession with a sample size of 50 cocoons /accession. Using the number of breaks during reeling cycle, non-broken filament length was computed and the data were statistically analysed using INDOSTAT software package. Cocoon shape variation was determined by uniformity test on the basis of standard deviation as suggested by Mano (1994). Based on the analysis of variance, standard deviation and co-efficient of variation were estimated for each accession to determine cocoon shape uniformity. Correlation was worked out between cocoon shape and size variables (average cocoon length, cocoon width and length-width ratio) with filament length and non-broken filament length. Germplasm accessions with less co-efficient of variation and standard deviations within the population were grouped.

RESULTS AND DISCUSSION

Based on the statistical analyses made, it was observed that significant variability existed among the 101 accessions for cocoon length, cocoon width, cocoon length-width ratio, total filament length and non-broken filament length (Table 5.1).

Correlation of cocoon length, cocoon width and cocoon length-width ratio with total filament length indicated non significant relationship values amongst them. Thus the results indicate that the cocoon length and cocoon width do not have any significant

Table 5.1: Variability in cocoon length, cocoon width, l/w ratio, filament length and non-broken filament length of silkworm (*Bombyx mori* L.) germplasm

Sl. No.	Accession No.	Race	Cocoon Length (mm)	Cocoon Width (mm)	Cocoon L/W Ratio	Filament Length (m)	Non-broken Filament Length (m)
1	BBE-0171	A25	33.12	19.62	1.688	1096.85	914.75
2	BBI-0172	Boropolu (Jammu)	34.18	20.11	1.700	1060.15	928.20
3	BBE-0173	CC (SL)	34.66	21.05	1.647	920.90	837.20
4	BBE-0174	Feng Shong	33.84	18.47	1.832	835.15	700.45
5	BBE-0175	Hong zhou (G)	31.23	19.77	1.580	1023.45	846.40
6	BBE-0176	Hong zhou (R)	32.52	20.55	1.582	896.65	666.40
7	BBE-0177	JPN5 x B25	36.91	18.09	2.040	991.65	941.90
8	BBE-0178	JPN5 x NK25	34.40	18.06	1.905	1020.70	887.30
9	BBE-0179	JPN6 x A26	33.41	20.89	1.599	1047.20	879.05
10	BBE-0180	JPN6 x B25	34.05	17.53	1.942	1052.11	911.05
11	BBE-0181	JPN12D	35.41	19.71	1.797	1086.70	934.05
12	BBE-0182	CSGRC-12	33.95	18.26	1.859	1061.90	970.10
13	BBE-0183	CSGRC-1	34.87	20.52	1.699	960.95	917.70
14	BBE-0184	CSGRC-2	33.32	20.43	1.631	1046.95	909.60
15	BBE-0185	CSGRC-3	32.64	15.67	2.083	1046.50	909.30
16	BBE-0186	CSGRC-4	34.25	19.03	1.800	1054.25	881.15
17	BBE-0187	CSGRC-5	33.85	18.66	1.814	1022.30	893.00
18	BBE-0188	CSGRC-6	36.56	17.94	2.038	1042.35	776.25
19	BBE-0189	Zebra Yellow	37.79	17.85	2.117	727.55	582.30
20	BBE-0190	36 PC	32.21	17.91	1.798	525.12	477.38
21	BBE-0191	39 P	36.50	19.96	1.829	703.43	633.08
22	BBE-0192	44 B M	34.11	17.28	1.974	1005.33	837.78
23	BBE-0193	44 F (M)	33.90	16.79	2.019	743.98	437.64
24	BBE-0194	644	35.02	17.56	1.994	668.00	401.05
25	BBE-0195	6p	32.14	20.67	1.555	981.96	818.30
26	BBE-0196	7042	31.19	19.58	1.593	901.06	901.06
29	BBE-0199	Auz-4	37.05	17.58	2.108	712.07	647.34
30	BBE-0200	Auz-5	32.09	16.31	1.968	744.94	622.55
31	BBE-0201	C124	34.19	21.95	1.558	961.08	524.22

positive or negative correlation with total filament length and non-broken filament length (Table 5.2), which predisposes the fact that cocoon length and width are not the contributing factors for filament length and breakage of filament during reeling process.

It was observed that 13 accessions viz., BBE 0175, BBE-0178, BBE-0179, BBE-0180, BBE-0181, BBE-0182, BBE-0185, BBE-0186, BBE-0187, BBE-0197, BBE-0205 and BBE-0270 showed more uniformity in cocoon shape as the standard deviation and co-efficient of variation were less than the overall mean values 1.88 and 8.00 respectively (Table 5.3). These accessions also showed the total filament length more than 1000 meters with SD and CV being 164.320 and 18.62 respectively. In all the accessions the average non-broken filament length was more than 825 metres with SD and CR being 193.545 and 27.46 respectively indicating less breakage during the total reeling process. Observations made on cocoon shape showed negative correlation (r = -.080) with reelability but those accessions having cither oval or elongated cocoon shape with faint constriction showed comparatively less filament breakage during reeling process and had long filament length of more than 1000 metres. It is evident from the results obtained that cocoon shape with deep constriction exerts more influence on filament breakage during reeling process and cocoon size variables like cocoon length and width do not contribute significantly for total filament length and non-broken filament length. Since mode of inheritance of cocoon shape in silkworm and the number of genes controlling the expression of cocoon shape and size have been established by Hirabayashi (1982) and Gamo *et al.*, (1985), the present study corroborate the earlier findings that cocoon shape along with those quantitative traits governing silk productivity needs to be given more emphasis for identifying or selecting suitable parents for breeding and evaluation programme to produce uniform shaped cocoons with desired filament characteristics.

Table 5.2: Correlation matrix of cocoon size variables with filament length and non-broken filament length in silkworm (*Bombyx mori* L.) germplasm

Parameter	Cocoon Length	Cocoon Width	Length Width Ratio	Filament Length	Non-broken Filament length
Cocoon Length	1.00000	-0.16973	0.58115	0.05350	0.07314
Cocoon Width	-	-	0.89617	0.00283	0.05973
Length Width Ratio	-	-	-	0.01338	-0.02327
Filament Length	-	-	-	-	.073855
Non-broken Filament length	-	-	-	-	-

Table 5.3: Statistics of best-ranked silkworm (*Bombyx mori* L.) germplasm accessions for uniformity in cocoon shape and filament characters

Accession No.	Cocoon length (mm)			Cocoon width (mm)			Length-width Ratio			Average Filament Len (m)	Average NBF (m)
	Mean	SD	CV%	Mean	SD	CV%	Mean	SD	CV%		
BBE-0175	31.23	1.590	5.090	19.77	0.726	3.672	157.967	9.145	5.785	1023.54	846.40
BBE-0178	34.40	1.664	4.836	18.06	1.047	5.800	190.476	12.699	6.650	1020.70	887.30
BBE-0179	33.41	1.498	4.484	20.89	0.861	4.123	159.933	4.906	3.067	1047.20	879.05
BBE-0180	34.05	1.682	4.942	17.53	1.088	6.211	194.238	15.412	7.905	1052.11	911.05
BBE-0181	35.41	1.516	4.280	19.71	0.883	4.484	179.655	10.062	5.592	1086.70	934.05
BBE-0182	33.95	1.489	4.386	18.26	0.779	4.267	185.926	11.148	5.986	1061.90	970.10
BBE-0185	32.64	1.550	4.749	15.67	0.742	4.739	208.296	9.861	4.730	1046.50	909.30
BBE-0186	34.25	1.462	4.270	19.03	0.763	4.011	179.979	8.387	4.655	1054.25	881.15
BBE-0187	33.85	1.417	4.187	18.66	0.786	4.215	181.404	9.955	5.480	1022.30	893.00
BBE-0192	34.11	1.260	3.694	17.28	0.958	5.549	197.396	9.994	5.053	1005.33	837.78
BBE-0197	33.40	1.435	4.297	19.04	1.077	5.658	175.420	10.583	6.018	1173.61	1043.21
BBI- 0205	35.48	1.643	4.632	20.86	1.026	4.921	170.086	9.402	5.519	1028.51	881.58
BBE-0270	34.76	1.548	4.455	17.86	0.905	5.076	194.625	7.429	3.814	1092.60	993.27

ACKNOWLEDGEMENTS

Authors thank Shri. R. Pugalendi, Technical Assistant for single cocoon reeling analysis and Shri. S. Sekar, Computer Programmer for statistical analysis of data.

REFERENCES

Chatterjee, S.N., Rao, P.R.M., Jayaswal, K.P., Singh, R. & Datta, R.K. (1993). Genetic variability in mulberry silkworm *Bombyx mori* L. breeds with low silk yield. *Indian J. Seric.*, 31(1): 69-86.

Gamo, T., Saito, S., Otusuka, Y., Hirobe, T. & Tazima, Y. (1985). Estimation of combining ability and genetic analysis by diallele crosses between regional races of the silkworm (2). Shape and size of cocoons. *Tech. Bull. Seric. Exp. Stn.*, 126: 121-135.

Hirabayashi, T. (1982). Influence of difference of cocoon shape in pure strains to that of hybrids. *Sansi-Kenkyu (Acta Sericologia)*, 121: 27-36.

Hiraishi, U. (1912). On the cocoon shape of hybrid in silkworm. *Dainihon-Sanshikaihou*, 21(251): 22-28.

Katsuki, K. & Nagasawa, S. (1917). Cocoon shape of silkworm hybrids. *Dainihon-Sanshikaihou*, 26(308): 8-15.

Lie, Z. (1996). Analysis of the inheritance stability of cocoon. *Abstr. Symposium on Sericulture Science facing 21st Century.* (6-10, Oct. 1996) p.105.

Mano, Y. (1994). Comprehensive Report on Silkworm Breeding. Central Silk Board, Bangalore, p.180.

Miura, M., Morikawa, H. & Sugiura, A. (1995). Statistical Analysis of body movement and shape of *Bombyx mori* L. in spinning behaviour. *J. Seric. Sci. Jpn.*, 64(3): 237-245.

Miura, M., Morikawa, H., Kato, H. & Iwasa, M. (1997). Analysis of the cocoon process of cocoon shape by *Bombyx mori*. *J.Seric. Sci. Jpn.* 66(1): 23-30.

Nakada, T. (1989). On the measurement of cocoon shape by use of image processing method, with an application to the sex-determination of silkworm *Bombyx mori* L. *Proc. of 6th International Congress*, pp. 957-960.

Nakada, T. (1991). On the sex linkage of cocoon shape in the silkworm. *Abstr. of 61st Cong. Jap. Soc. Seric. Sci.*, p. 50.

Nakada, T. (1992a). Sequence of some cocoon traits in the progeny tests after crossing between wild and domesticated silkworm. In: *Wild Silkmoths*, Akai. H, Y. Kato, M. Kiuchi and J. Kobayashi (eds.). pp. 98-104.

Nakada, T. ((1992b). The principal component analysis regarding cocoon shape in the silkworm with special reference to the difference between male and female. *Abstr. 62nd Cong. Jpn. Soc. Seric. Sci.*, p. 92.

Nakada, T. (1993). Genetic differentiation of cocoon shape in the silkworm, *Bombyx mori* L. *Int. Cong. Genetics*, p. 224.

Nakada, T. (1994). On the cocoon shape measurement and its statistical analysis in the silkworm, *Bombyx mori*, L. *Indian J. Seric.*, 33(1): 100-102.

Nakada, T. (1998). A statistical analysis on the genetic differentiation of cocoon shape in the silkworm, *Bombyx mori* L. *Memoirs Fac. Agric. Hokkaido Univ. Japan*,. 21(1): 101-109.

Nakada, T., Maeda, H. & Murakami, M. (1991). Discriminant analysis related to sex difference of cocoon shape in the silkworm, *Bombyx mori* L. *Proc. Inst. Stat. Mathematics*: 36(1), 23-40.

Rao, P R.M., Premalatha, V., Singh, R. & Vijayaraghavan, K. (1997). Variability studies in some pure races and F1 hybrids of the silkworm, *Bombyx mori* L. *Environ. Eco.*, 15(3): 683-687.

PERFORMANCE OF REELING TRAITS IN MULBERRY SILKWORM HYBRIDS

A.A. Siddiqui, R.K. Khatri, Deepak Kumar, D.P. Srivastava and Meera Verma*

ABSTRACT

The reeling parameters of cocoons viz. shell weight silk ratio, silk weight, waste weight, filament length, reelability percentage, denier and renditta are having paramount importance to produce quality raw silk. In present paper, reeling performance of four newly evolved hybrids was studied. Besides, inter-relationship of various reeling traits with absolute silk yield and among them were worked out. A strong association (+0.8301**, +0.8665**) was observed between absolute silk yield and shell weight & shell ratio. Similarly, strong association was also found between silk ratio and silk weight, filament length, reelability percentage and denier. Silk weight too was associated with filament length (+0.9040**) and denier (+0.6566**). Likewise, reelability percentage was strongly associated with renditta (+0.7649**).

The above studies suggest that traits like silk weight, reelability percentage, denier, filament length etc. can be improved further through breeding.

Key words: Reeling parameters, Hybrids, Denier, Filament length, Inter-relationship.

INTRODUCTION

Good quality cocoons should be compact in built, uniform in shape and size, less flossy and easily reelable. These parameters are not only determined by the Silkworm races but the care taken at the time of spinning of cocoons in terms of required environmental

Regional Development Office, Central Silk Board, 5th Floor, "Vikas Deep", Lucknow – 226 001

*Zonal Silkworm Seed Organization, Majra, Dehradun – 248 171

E-mails: zzsspoddn@sancharnet.in; guptadk.deepak@gmail.com.; siddiquibad@yahoo.co.in conditions play also a very vital role. Interrelationship between traits provides great help to the silkworm breeders in selection of suitable breeding lines besides, providing information for the best way to handle segregating population to combine desirable traits while eliminating undesirable ones. In mulberry silkworm correlation between traits have been studied by many investigators (Hirobe, 1967; Ohi *et al.*, 1970; Yokoyama, 1979; Rajanna and Reddy, 1990; Eguchi *et al.*, 1995; Naseema Begum and Yamamoto, 2002 and Nanje Gowda *et al.*, (2003). Hence, the present study was undertaken to see the performance of reeling parameters and also to find out the correlation coefficient (r) among the reeling traits in new bivoltine hybrids viz a viz conventional NB4D2 × SH6 as control.

MATERIALS AND METHODS

The rearing of four bivoltine hybrids viz CSR2 × CSR4, Dun6 × Dun7, RSJ14 × RSJ11 and NB4D2 × SH6 was conducted at DOS farm Dehradun. 300 cocoons per replication were subjected to reeling studies for three replications per hybrid. The reeling tests were performed on standard instruments following the standard methods.

The traits related to breeding point of view was recorded as: Absolute Silk yield (g), Shell weight (g), Shell ratio (%), Silk weight (g), Waste weight (g), Unbreakable filament length (m), Reelability (%), Denier (No) and Renditta (Kg).

The data of traits so recorded were statistically analyzed alongwith correlation matrix following the procedure of Singh and Choudhary, (1985).

RESULTS AND DISCUSSION

The reeling traits and its correlation in four bivoltine hybrids in Tables 6.2 & 6.3. The perusal of the data reveal that absolute silk yield, shell weight, shell ratio percentage, silk weight, waste weight, filament length, reelability percentage, denier and renditta were recorded 235.44, in Dun 6 × Dun 7, 0.36 kg. in Dun 6 × Dun 7, 92.24% in CSR2 × CSR4, 91.90 in NB4D2 × SH6, 8.20 in NB4D2 × SH6, 883 m in NB4D2 × SH6, 87.0% in RSJ14 × RSJ11, 2.70 in RSJ14 × RSJ11, 7.30 kg in CSR2 × CSR4. This is also evident from the table that since the reelers are accustomed to conventional bivoltine hybrid, unbreadable filament length is more and waste weight is less in NB4D2 × SH6. The wide range traits also indicate significant difference within the hybrids and their genetic make up.

The ultimate goal of silkworm breeding programme is to evolve high yielding hybrids for commercial exploitation which is largely depended on the choice of parents,and breeding plans followed by appropriate selection procedures. To overcome this, the breeder should know the inheritance of various traits, response to selection and relationships among the traits.

The data presented in Table 6.2 clearly indicate that majority of traits are correlated with each other either positively or negatively. The trait reliability is mainly influenced by environmental factors such as temperature, humidity and aeration. For improvement in silkworm breed, breeders are required to keep in mind all these correlated traits (Table 6.1).

Studies in the past have shown that selection for one trait has correlation with other traits (Kobari and Fujimoto, 1966; Ohi *et al.*, 1970; Miyahara, 1978; Eguchi *et al.*, 1995). Ohi *et al.*, 1970 reported that in silkworm many traits such as cocoon weight, shell weight, shell ratio, filament length and filament weight are having direct relationship with quantity and quality of silk yield which is the main objective of silkworm rearing. Nagaraju (1984) reported that filament length was decreased when selection was carried out for high shell weight. Whereas, Kurasawa (1968b) found positive correlation of the shell weight and shell ratio with filament length, the results obtained in present study also confirmed the results of Kurasawa (1968b).

Table 6.1: The association recorded between the reeling traits.

Name of the Hybrids	Absolute Silk yield (g)	Shell weight (g)	Shell ratio (%)	Silk weight (g)	Waste weight (g)	Unbreakable Filament length (m)	Reelability (%)	Denier (No)	Renditta (Kg)
CSR2XCDR 4	180.25	0.35	19.24	87.60	12.40	839.0	85.00	2.50	7.30
DUN6XDUN 7	235.44	0.36	18.46	87.50	12.50	851.0	85.00	2.40	8.00
RSJ14XRSJ 11	147.25	0.25	19.23	89.30	10.70	873.0	87.00	2.76	7.80
NB4D2XSH6	189.42	0.33	18.54	91.90	8.20	883.0	81.00	2.70	8.50
CD at 5%	46.81	0.10	0.31	0.80	0.80	2.61	0.89	0.22	0.40
CD at 1%	66.58	0.13	0.42	1.15	1.14	3.72	1.27	0.30	0.57

Table 6.2: Performance of reeling traits in four bivoltine Silkworm hybrids.

Character	Shell weight (g)	Shell ratio (%)	Silk weight (g)	Waste weight (g)	Unbreakable Filament Length (m)	Reelabilty (%)	Denier (No)	Rendiitta (Kg)	Absolute Silk Yield (g)
Shell weight (g)	1.000	0.58882	+ 0.5806	+ 0.6145	+ 0.5633	- 0.2539	- 0.7229**	+ 0.5478*	0.8301**
Shell ratio (%)		1.000	+ 0.7788**	0.7606**	+ 0.7385**	+ 0.7177**	+ 0.6681*	0.0697	+ 0.8665**
Silk weight (g)			1.000	- 0.9988**	+ 0.9040**	- 0.6959*	+ 0.6566*	- 0.6492*	+ 0.3044
Waste weight (g)				1.000	+0.9153**	- 0.6910*	- 0.4926	0.6482	+ 0.3126
Unbreakable Filament length (m)					1.000	- 0.4148	+ 0.8366**	+ 0.6833*	+ 0.3474
Reelabilty (%)						1.000	- 0.2179	+ 0.7649**	+ 0.6334*
Denier (No)							1.000	- 0.4676	- 0.7632**
Randitta (Kg)								1.000	- 0.2423

* Significant at 5% level; ** Significant at 1% level.

Table 6.3: Correlation coefficient of Absolute Silk yield and other reeling traits in mulberry silkworm hybrids.

S. No	Traits	Positive correlation	Negative correlation
1	Absolute Silk yield (g)	Shell weight ** Shell ratio ** Silk weight Waste weight Un. Filament length * Reelability ratio** Denier Renditta	
2	Shell weight (g)	Shell ratio ** Silk weight Waste weight Un.Filament length * Renditta *	Reelability Denier**
3	Shell ratio (%)	Silk weight ** Waste weight ** Un.Filament length ** Reelability Denier	
4	Silk weight (g)	Un.Filament length ** Denier **	Reditta Waste weight ** Reelability** Renditta
5	Waste weight (g)		Un.Filament ** Reelability* Denier* Renditta
6	Unbreak. Filament length (m)	Denier Renditta	Reelability
7	Reelability (%)	Renditta	Denier
8	Denier (No)		Renditta

Based on correlation coefficient values following observations were made. Majority of traits like shell weight, shell ratio, filament length and denier are correlated with each other and with many other traits. The tendency of +ve or –ve correlations for different traits can be exploited by the application of systematic selection and bring together some advantageous features of the parental breeds. Therefore, in silkworm, one needs to be quite careful while selecting different traits simultaneously.

REFERENCES

Eguchi, R., Shimazaki, A., Ichiba, M. & Shibukawa, A. (1995). Breeding of the high yielding silkworm races "Shoho" (NO2 × CO2) and "Ohwaski" (N150 × C150). *Bull. Natl. Inst. Seric. Intomol. Sci.,* 12: 47-49.

Hirobe, T. (1967). Advancement in the improvement of Silkworm varieties. *Heredity,* 21: 18-24.

Kobari K. & Fujimoto, N. (1966). Studies on the selection of cocoon filament length and cocoon filament size in *Bomby muri,* L. *Nissenzatsu,* 35(b): 427-434.

Kurasawa, H. (1968b). Selection of quantitative cocoon characters in the silkworm II. Changing of cocoon characters with selection of weight of cocoon layer and cocoon layer ratio. *J. Seric. Sci. Jpn.,* 37(1): 51-56.

Miyahara, T. (1978). Selection for long filament length basic variety, MK effect of selection in the later generation. *Acta. Senicilogia,* 106: 73-78.

Nagaraju, J. (1984). Silk yield attributes. In: *Correlation and Complexities in Silkworm Breeding,* G. Sreerama Reddy *et al.,* (Eds). Oxford and IBH Pub. Co. Pvt. Ltd., New Delhi, Calcutta, pp. 168-185.

Gowda, N.B., Redday, M.N., Paliti, A.K., Kumar, S.N. & Kalpana, G.V. (2003). Correlation of various qualitative and quantitative characters in Silkworm. *Bomby mori,* L. Advances in Tropical Sericulture. *Proc. Nat. Conf. on tropical Sericulture for global competitiveness CSR & TL,* Mysore, p. 82-87.

Begum, N.A. & Yamamoto, T. (2002). Correlation coefficient studies on certain quantitative traits in the silkworm *Bomby mori* L. *Int. J. Indus Entom.,* 1: 45-52.

Ohi, H., Miyahara, J. & Yamashita, A. (1970). Analysis of various practically important characteristics in the silkworm in the early breeding generation between parents and offspring as well as relation between each character. *Tech. Bull. Sericult. Expt. Sta.* MAFF, 93: 39-49.

Rajanna, G. S. & Reddy, R.G. (1990). Studies on the variability and inter relationship between some quantitative characters in different breeds of silkworm *Bombyx mori* L. *Sericologia*, 30(1): 67-73.

Singh, R. K. & Chaudhary, B.D. (1985). *Biometrical Methods in quantitative Genetic Analysis*, Kalyani Publishers, Ludhiana. pp. 39-53.

Yokoyama, T. (1979). Silkworm selection and hybridization In: *Genetics in Relation to Insect Management* Working papers. The Rockfeller Foundation Management. pp. 71-83.

7

BIODIVERSITY OF THE MULBERRY SILKWORM *BOMBYX MORI* L.

Geetha N. Murthy, H.S. Phaniraj and B. Saratchandra

ABSTRACT

During recent years. biodiversity conservation programmes have drawn the attention of many countries including developing nations, because of the genetic erosion due to indiscriminate use of bio resources and damage to the environment. destruction of forest land, human interference in eco-system upsetting the equilibrium of the biosphere. In spite of industrial development resulting in de-forestation, India still can claim as the owner of wide bio-diversity with a rich wealth of sericigenous flora and fauna distributed abundantly especially in northern India for silkworm spp. and in the North Eastern region, Sub-Himalayan ranges in North Western part of India and also in the Western Ghats.

In India, the Central Silk Board is the main organization involved in the conservation of silkworm germplasm. In 1990, the Govt. of India established a sericultural germplasm centre viz. The Central Sericultural Germplasm Resources Centre (CSGRC) at Hosur under the National Sericulture Project, exclusively for maintenance of sericulture germplasm. Presently, CSGRC Hosur maintains more than 350 silkworm, germplasm resources and catalogues are available for 628 mulberry and 335 silkworm, germplasm and has become one of the largest repositories of sericultural germplasm in the world. In addition, the centre also has a continuous plan of action to collect new silkworm germplasm resources and supplies silkworm germplasm for crop improvement programmes.

CSGRC. Hosur, has developed a user friendly software system, the "Silkworm Germplasm Information System (SGIS)" for data recording,

Central Silk Board, Ministry of Textiles, Govt. of India, C.S.B. Complex, Bangalore-560068, Karnataka.
E-mail: sarat@silkboard.org.

data retrieval and data mining and a catalogue on silkworm germplasm has also been published (Sekar and Thangavelu, 2003). This user friendly software system developed by CSGRC Hosur can be further developed to create a single system of database maintenance which can be accessed by all countries involved in sericulture for better information exchange and utility. Proper documentation on the availability of silkworm germplasm in different countries is absolutely necessary (Thangavelu, 2002). It is very essential to conserve and utilize the wild as well as domesticated resources of mulberry silkworm *Bombyx mori* as natural selection induced genetic diversity is rather very limited to voltinism. It is essential to share the information on silkworm germplasm among the countries involved in sericulture as well to exchange available silkworm germplasm among the countries for the benefit of mankind in the world which in turn will promote silkworm breeding and sericulture development. At the National level, it is imperative that, a network be established among all the institutes/ universities/organizations including CSB within the country so that, a system for effective conservation of the mulberry silkworm biodiversity at regional and thereby at national level is developed. At the International level, the FAO can play an important role to promote the programme of silkworm breeding and also facilitate need based exchange of sericultural germplasm among the member countries involved in sericulture. The FAO can also help to develop a global sericulture development strategy including global market information service for international silk trade. Once, such a system is developed and implemented effectively for *Bombyx mori*, a similar line of approach may be extended for non-mulberry silkworms also initially through establishment of regional centres for their conservation in regions where non-mulberry silkworms are exclusively reared.

Keywords: Biodiversity, *Bombyx mori*, CSGRC Hosur, Germplasm maintenance, FAO

INTRODUCTION

Biodiversity is the variety of all forms of life on earth and its complexity is in terms of variations at genetic, species and ecosystem levels. Broadly speaking, the term 'biodiversity' encompasses all species of plants, animals, micro-organisms, their genetic materials and the ecosystems of which they are part. Biodiversity is the result of evolution that is a continuous phenomenon induced by natural selection and as a result the population of organisms evolves through adaptation to the biotic and abiotic stress. In recent years, the loss of entire species and natural areas has been occurring at unprecedented rates.

Biodiversity is required for the recycling of essential elements like carbon, oxygen and nitrogen, for mitigating pollution, protecting watersheds, combating soil erosion and to act as a buffer against excessive variations in weather and climate for protection from catastrophic events beyond human control (McNeely, 1990). The importance of biodiversity for a healthy environment has become increasingly clear and if we over exploit living resources, we threaten our own survival (Anonymous, 1992).

The economic value of biodiversity is a well-established fact. Modern agriculture depends on new genetic stock from natural ecological systems and nature tourism generates billions of annual revenue worldwide (Anonymous, 1992). Breeders and farmers rely on the genetic diversity of crops and livestock to increase yields and to respond to changes in environmental conditions. Thus the genetic biodiversity protects both farmers and consumers alike. The challenge of biodiversity research entails not only the gathering of information, but its management, application, and communication (Anonymous, 1992).

CURRENT SCENARIO

During recent years, biodiversity conservation programmes have drawn the attention of many countries including developing nations, because of the genetic erosion due to indiscriminate use of bio resources and damage to the environment, destruction of forest land, human interference in eco-system upsetting the equilibrium of the biosphere. Thus the concept of biodiversity conservation and gene bank maintenance has gained greater importance and momentum and the biodiversity wealth is being increasingly considered as common heritage of mankind/right of the nation. The issues related to accessing genetic resources, its sustainable use, benefit sharing, rights of farmers etc. are being deliberated at various national and international forum. Realizing the importance of biodiversity conservation for sustainable development of agriculture, the Consultative Group on International Agricultural Research (CGIAR) established the International Board for Plant Genetic Resources (IBPGR) in 1974 at Rome with a global network of genetic resource centres, mainly for conservation of natural genetic resources including the wild species to promote crop improvement programmes and increase food production (Srivastava & Thangavelu, 2005).

It is well established that the heritage of usage of silk for dress materials in India, Russia, and China dates back to pre-medieval period. In spite of industrial development resulting in de-forestation, India still can claim as the owner of wide bio-diversity with a rich wealth of sericigenous flora and fauna distributed abundantly especially in northern India for silkworm spp. and in the North Eastern region, Sub-Himalayan ranges in North Western part of India and also in the Western Ghats. Silkworm rearing was prevalent in Kashmir and North Eastern states during sixteenth century, the Mughal period where the univoltine and multivoltine silkworms were respectively reared and the Tippu Sultan introduced silkworm rearing in south India in 1875. During eighteenth century, the British rule in India, quite a few univoltine and bivoltine races were imported from Italy, France, Russia and China, and the races were bred and maintained by the farmers (Krishna Rao, 1997); and there was no systematic maintenance of the silkworm germplasm and hence only few races survived under Indian climatic condition.

At present, only few old indigenous races are surviving viz. Barapolu, Chotapolu, Nistari, Sarupat, and Moria, whereas the indigenous univoltine Kashmiri races are almost extinct. In India, systematic silkworm stock maintenance and breeding started in the early nineteenth century. Prior to 1922, only pure races were reared and hybrid silkworms were introduced later, Pure Mysore

× C. Nichi was probably the first hybrid in Karnataka and exploitation of hybrids in West Bengal and Kashmir came much later during 1956 and 1959 respectively (Thangavelu, 1997). Silkworm genetic stock maintenance started during 1940 in an organised way at Sericultural Research Station, Berhampore in West Bengal and subsequently temperate silkworm germplasm stocks were established at Univoltine Silkworm Seed Station, Pampore in Kashmir and multivoltine and bivoltine silkworm stocks were established at Central Sericultural Research and Training Institute, Mysore in Karnataka and Coonoor in Tamil Nadu (Thangavelu, 2002).

PRESENT STATUS OF MAINTENANCE/ CONSERVATION OF SILKWORM GERMPLASM RESOURCES

The sericulture industry depends on the rich silkworm gene pool and sericulture advancement is largely based on the silkworm variety improvement as well as breeding of new silkworm variety which in turn is based on the supply of rich silkworm germplasm resources. These resources are the diversity of germplasm that could always supplement the new genetic resources for the breeders. Thus, sericulture is supported by the silkworm gene resource diversity (Kee-Wook Sohn, 2003). Many genetic discoveries in silkworm belong to the forefront of genetic research and they enriched the genetic theory development. Maintenance of silkworm germplasm resources is quite different from other agricultural gene bank and can be artificially maintained indoor.

Since the discovery of heterosis in silkworm breeding, hybrids are used almost all over the world in commercial cocoon production, which has greatly enhanced both in quantity and quality of cocoon yield per unit. However, the tendency of similarity in main economic characters and close genetic relationship has resulted in the elimination of vast local varieties with significant ecological diversity and loss of some specific genes. Therefore, ecological adaptability of the improved varieties greatly decreased causing genetic-erosion. One of the evidence is the decay of sericulture in Europe in 19th century due to epidemic of the pebrine disease which ruined European sericulture in view of the pebrine-sensitivity of silkworm varieties used during that period. Sustainable development of sericulture industry is largely depending on breeding new silkworm varieties with one or more distinctive character by effective utilization of silkworm germplasm resources. Silkworm germplasm resource is the necessary material base for sustainable development of sericulture (Kee-Wook Sohn, 2002).

In India, the Central Silk Board is the main organization involved in the conservation of silkworm germplasm. Earlier, the conservation of silkworm germplasm was carried out by its Research Institutes viz. Central Sericultural Research & Training Institutes (CSR&TI) at Berhampore, Mysore and Pampore, Regional Sericultural Research Stations (RSRS) at Jammu, Dehradun, Kalimpong and Coonoor; some of the universities and state funded sericultural research institutes. In 1990, the Govt. of India established a sericultural germplasm centre viz. The Central Sericultural Germplasm Resources Centre (CSGRC) at Hosur, under the National Sericulture Project, exclusively for maintenance of sericulture germplasm. After the establishment of the CSGRC, the entire base collection of silkworm germplasm was collected from these institutes and universities and maintained in the national repository for further conservation, maintenance and utilization. The silkworm germplasm thus maintained are utilized to evolve breeds, rejuvenate the popular breeds and identification of potential donor from the available genetic resources through evaluation and selection. The aforesaid institutes/regional stations of CSB as well as the universities/State research institutes do continue to maintain the silkworm germplasm available with them for utilizing in the silkworm breeding experiments.

Presently, CSGRC Hosur maintains more than 350 silkworm germplasm resources and catalogues are available for 628 mulberry and 335 silkworm germplasm. Thus CSGRC, Hosur, has become one of the largest repositories of sericultural germplasm in the world. In addition, the Centre also has a continuous plan of action to collect new silkworm germplasm resources through survey and exploratory programmes. Based on the indents received from Research Institutes and Universities, the Centre supplies silkworm germplasm for crop improvement programmes and the supply is regulated through a Germplasm Supply Committee constituted for this purpose. Only those silkworm germplasm authorized by the National Race Authorization Committee constituted by the CSB are supplied. Indents for silkworm accessions outside the list of authorized accessions are also screened by the Germplasm Supply Committee and based on the merits of the indents and specific study, they are supplied. The request from other countries for supply of silkworm germplasm is decided by the Member Secretary, Central Silk Board, Govt. of India on merit basis (Srivastava & Thangavelu, 2005).

CO-OPERATION THROUGH EXCHANGE/SUPPLY OF SILKWORM GERMPLASM RESOURCES

In sericulture, although, national co-operation is very much in place through exchange/supply of silkworm germplasm, international co-operation is very much limited. However, countries like Japan, South Korea, China and India extend financial/technical/training support in sericulture on specific request from other countries and the silkworm germplasm are exchanged to a limited extent under bilateral agreements/mutual co-operation. India has extended technical co-operation in sericulture to Sri Lanka under Indo-Colombo Plan and South Asian Association for Regional Co-operation (SAARC). Inspite of the above, the exchange of sericultural germplasm is very much restricted at the global level among the countries and that too only through diplomatic channel or personal contacts. Thus, there is no need based and regular exchange of silkworm germplasm among the countries for silkworm breeding and other research purpose as in the case of agriculture, horticulture, animal husbandry etc.

STATUS OF DATABASE DOCUMENTATION

Presently, CSGRC, Hosur, is the exclusive nodal organisation for collection and conservation of seri genetic resources and also serves as a national data base documentation centre for sericultural germplasm. The centre has developed a user friendly software system with data entry screen, creation and maintenance of files, data pooling, data base querying, data generation for statistical analysis etc., in Foxpro designated as "Silkworm Germplasm Information System (SGIS)" for data recording, data retrieval and data mining. A catalogue on silkworm germplasm has also been published with exhaustive data on characterization and evaluation, utilizing the software system developed by the centre for future studies on germplasm and to facilitate monitoring of conservation process (Sekar and Thangavelu, 2003).

NEW VISTAS EXPLORED

A new vista has been opened to understand nature's bio-diversity and its relevance through adoption of molecular tools. The same approach is being availed and research is in progress for building up the foundation work on the bio-diversity of silkworm species in India under CSB. (Chatterjee & Tanushree, 2004).

FUTURE STRATEGY

DATABASE MAINTENANCE

In view of the rich wealth of sericigenous flora and fauna available in India, it is imperative that the conservation of these genetic resources under *ex-situ* conditions and effective utilization on sustainable basis have to go hand in hand so that, the usefulness of large silkworm germplasm collections can be enhanced to justify long-term investments on conservation of gene pool. In order to sustain the utilization of these genetic resources, proper maintenance of database collected over several generations on their subjective and objective descriptors is the need of the time (Sekar and Thangavelu, 2003). CSGRC, Hosur, has developed a user friendly software system with data entry screen, creation and maintenance of files, data pooling, database querying, data generation for statistical analysis etc. This can be further developed to create a single system of database maintenance which can be accessed by all countries involved in sericulture for better information exchange and utility.

DOCUMENTATION

There are approximately 4310 silkworm germplasm accessions available in various countries. Many of the countries where sericulture is on the decline/has declined completely, have maintained the available silkworm germplasm resources in their research institutes and hence, have their own special gene banks. In such countries, in case of failure in any of the institutes, the silkworm germplasm resources will disappear from the world. Therefore, it would be very useful if an international cooperation network is established for joint maintenance as well as exchange of silkworm germplasm resources. Further, there is always a likelihood of some of the accessions available with the countries being duplicated. Hence, a proper documentation on the availability of silkworm germplasm in different countries is absolutely necessary (Thangavelu, 2002).

GERMPLASM MAINTENANCE

As regards maintenance of silkworm germplasm resource, the aim should be to stablize filial generation; maintain/store the mutation strains; provide details on the genetic mechanism of gene information and expression; study new methods and techniques for germplasm maintenance, management and multiplication; exchange and manage the germplasm information system and keep certain DNA gene fragment and study the gene/DNA structure and function and to exploit new materials.

GERMPLASM CONSERVATION

The mulberry silkworm *Bombyx mori* being domesticated, natural selection induced genetic diversity is rather very limited to voltinism. Hence, it is very essential to conserve and utilize the wild as well as domesticated resources of *Bombyx mori* to broaden its genetic diversity, apart from the geographical races, mutants, sex-limited races, evolved breeds and genetic stocks of breeders. The wild relatives of *Bombyx* are very vulnerable, but the extent of vulnerability at different scales is not yet known. The design of biodiversity network in sericulture involving wild relatives and domesticated *B. mori* is also not well established. Therefore, conservation of wild as well as domesticated resources is very essential for sustainable development of sericulture since loss of genetic resources of domesticated and wild relatives of *Bombyx* species along with their unique genes will definitely prove to be a disadvantage for the future generation.

In agricultural, horticultural and sericultural crop improvement programme, the wild species of several crop plants have contributed very valuable genes for resistance to diseases and pests and tolerance to adverse agroclimatic conditions [Jackson and Ford-Lloyd, 1990]. However, similar exploitation of genes from wild relatives of *B.mori* has not been reported. The role of wild relatives and wild species in agricultural crop improvement are well known. Improvement in silkworm race heavily depended on the geographical races of *B. mori* and the wild relatives of *Bombyx* were not explored, unlike in agriculture. Hence, there is an urgent need for conservation of the wild relatives of *Bombyx* and Bombycidae.

INFORMATION SHARING

Sericulture is practiced both in developing and under developed countries benefiting several million people mainly from rural areas and economically weaker sections of the society. Therefore, development of sericulture will definitely improve the quality of life of economically poor people in these countries. In this direction, the need of the hour is robust silkworm breeds, and breeds tolerant to wider ranges of agro-climatic conditions. However, it is observed that, the information sharing on silkworm germplasm at global level among the countries is very much limited, unlike in agriculture, horticulture, sericulture, animal husbandry and fisheries etc. Further, in several countries sericulture is not sustainable and cost of production of silk is very high mainly because suitable silkworm breeds are not available and they are also unable to evolve silkworm breeds suitable to their own agro-climatic conditions. Therefore, it is essential to share the information on silkworm germplasm among the countries involved in sericulture as well to exchange available silkworm germplasm among the countries for the benefit of mankind in the world which in turn will promote silkworm breeding and sericulture development.

India maintains both exotic and indigenous mulberry and silkworm germplasm. To maximize high cocoon productivity, there is a need to collect higher cocoon and leaf yielding genotypes of silkworm from sericulturally advanced countries besides exploiting hitherto unexploited indigenous genotypes for breeding/genetic improvement programme (Chinnaswamy, 2001).

SUPPORT FROM EXTERNAL AGENCIES

At National Level

It is well-known that, within the country several institutes/universities/organizations other than those functioning under CSB also maintain silkworm germplasm collections for conducting various experiments/development schemes. However, a concrete network on the availability/details on these collections like region-specific, disease tolerant/resistant/hardy germplasm resources capable of surviving under adverse climatic conditions etc. are not in place. Hence, it is imperative that, a network be established among all the institutes/universities/organizations including CSB within the country so that, a system for effective conservation of the mulberry silkworm biodiversity at regional and thereby at national level is developed. This would in turn, aid in paving the way for taking up initiatives to formulate an international network.

At International Level

The Food and Agriculture Organization of the United Nations (FAO) acts as a neutral forum where all nations meet and provide help to developing countries and countries in transition to modernize and improve in the fields of agriculture, forestry, fisheries practices etc. with special focus on developing rural areas. In view of this role, the FAO can play an important role to promote the programme of silkworm breeding and also facilitate need based exchange of sericultural germplasm among the member countries involved in sericulture. The silkworm and mulberry germplasm available in countries where sericulture has declined may be deposited in China/India where sericulture is still thriving through financial assistance and infrastructure facilities from FAO, similar to the International Agricultural Research Centres functioning under the auspices of CGIAR and facilitate exchange as well as conservation and utilization of sericultural germplasm among the FAO member countries similar to International Board for Plant Genetic Resources (IBPGR), Rome. An international centre be established for conservation of sericultural germplasm and formulate the procedure, terms and conditions for the exchange of germplasm among the FAO member countries.

It is very essential to develope a global market information service for international silk trade covering market demand - country wise, fashion demand - season wise, along with other information on clothing behaviour of people from different countries and competition from other textiles, which will help silk producing countries to regulate their silk production, exports and imports of silk yarn, fabrics, dress materials etc. The FAO can help to develop a global sericulture development strategy including global market information service for international silk trade. The stability and positive growth of world silk trade has to be ensured so that utilization of sericultural germplasm among the FAO member countries is promoted through sericulture germplasm resources management, silkworm disease control and other programmes.

Thus, development of a global market information service for international silk trade will help a lot in reducing the adverse effects of the often observed fluctuations in international silk trade which is mainly due to lack of information on global market trends, market demand, price factors etc. Thus the negative influence on the domestic market and ultimately on the very root of sericulture can be minimized. To promote sericulture, a potential production system for sustainability of agro-ecosystem among SAARC countries, India needs to play a major role to formulate policies and programmes through an organized regional network in the Asian region (Chinnaswamy, 2001).

Thus, a global system needs to be developed for effective conservation of the mulberry silkworm biodiversity. Once, such a system is developed and implemented effectively for *Bombyx mori*, a similar line of approach may be extended for non-mulberry silkworms also, initially through establishment of regional centres for their conservation in regions where non-mulberry silkworms are exclusively reared.

REFERENCES & SUGGESTED READINGS

Anonymous (1992). Conserving Biodiversity: *A Research Agenda for Development Agencies.* Washington, National Research Council, D.C., National Academy Press, p. 118.

Anonymous (1992a). Global Biodiversity Strategy. The World Conservation Union and United Nations Environment Program, Washington, D.C. p. 262.

Sohn, B.H. (2002). Conservation Status of Silkworm Genetic Resources in Korea. Paper presented in the satellite session of *19th Congress of International Sericulture Commission*, Bangkok, Thailand, 21-25 September, 2002.

Kyushu, B. (2002). Conservation Status of Silkworm Genetic Resources in Japan. Paper presented in the Satellite session of 19th Congress of International Sericulture Commission, Bangkok, Thailand, 21-25 September, 2002.

Chatterjee, S. & Tanushree (2004). Molecular profiling of silkworm biodiversity in India: An overview. *Russian Journal of Genetics,* 40 (12): 1339 - 1347(9).

Chinnaswamy, K.P. (2001). Seri-biodiversity in India. *Proceedings of Regional Network Seminar and 2nd General Assembly on participatory biodiversity conservation in the South Asia Region,* 10 – 11, Feb.2001, Kathmandu, Nepal, pp. 54-61.

Hochberg, M.E. (2000). What, Conserve Parasitoids? In: *Parasitoid Population Biology.* Princeton. Eds., M. E. Hochberg and A. R. Ives, University Press, New Jersey. Chapter 17. p. 366.

Jackson, M. & Ford-Lloyd, B. (1990). Plant genetic resources - a perspective. In: *Climatic Change and Plant Genetic Resources.* (eds. Jackson & Ford-Lloyd). London, Belhaven Press, pp. 1-17.

Sohn, K.W. (2002). Conservation Status of Sericulture Germplasm (*Bombyx mori*) Genetic Resources in the World - Paper presented at *the Satellite session of 19th Congress of International Sericulture Commission,* Bangkok, Thailand, 21-25 September, 2002.

Kosegawa, E. (2002). Conservation Status of Silkworm Genetic Resources in Japan. Paper presented in *the Satellite session of 19th Congress of International Sericulture Commission*, Bangkok, Thailand, 21-25 September, 2002.

Rao, K.S. (1997). History of silkworm races. In: *Silkworm Breeding.* Edt. G.. Sreerama Reddy. Oxford & IBH Publishing. Co. Pvt. Ltd. New Delhi, 3-17.

McNeely, J., Miller, K.R., Reid, W.V., Mittermeier, R.A., Werner, T.B. (1990). *Conserving the World's Biological Diversity.* Washington, D.C., IUCN, WRI, CI, WWF, and The World Bank. pp. 416-417.

Sekar, S. & Thangavelu, K. (2003). Seribiodiversity database and management – Paper presented in the National Seminar on "*Bioinformatics and Biodiversity Data Management*" at Tropical Botanic Garden and Research Institute (TBGRI), Palode, Thiruvananthapuram from 15th to 17th May, 2003.

Srivastav, P.K. & Thangavelu, K. (2005). *Sericulture and Seri-Biodiversity.* New Delhi, Associated Publishing Company. p. 254 .

Thangavelu, K. (1997). Silkworm breeding in India - An overview. *Indian Silk,* 36(2): 5-13.

Thangavelu, K. (2002). Conservation Status of Silkworm Genetic Resources in India. Paper presented at the *Satellite session of 19th Congress of International Sericulture Commission*, Bangkok, Thailand, 21-25 September, 2002.

Tzenov, P.I. (2002). Conservation Status of Silkworm Genetic Resources in Bulgaria. Paper presented at the *Satellite session of 19th Congress of International Sericulture Commission*, Bangkok, Thailand, 21-25 September, 2002.

Windsor, D.A. (1995a). Equal Rights for Parasites. *Conservation Biology,* 9(1): 1-2.

Windsor, D.A. (1995b). Endangered Interrelationships: The Ecological Cost of Parasites Lost. *Wild Earth*, 5(4): 78-83.

PERFORMANCE OF DIFFERENT MULBERRY VARIETIES ON SILKWORM REARING IN MONSOON SEASON OF UTTAR PRADESH

ANSHUL SRIVASTAVA AND VADAMALAI ELANGOVAN

ABSTRACT

An investigation was carried out to asses the feeding quality of mulberry varieties (AR-10, AR-12 and AR-14) over Chak Majra (control) by silkworm rearing in monsoon season of Uttar Pradesh climate. In general, the three mulberry varieties were found to be superior over Chak majra as assessed by the rearing performance. However, the best rearing performance was observed from AR-10 during monsoon season of Uttar Pradesh. In addition, the study documents the effect of different mulberry varieties on larval growth and duration, ERR, shell weight, cocoon weight and shell ratio percentage.

Key words: *Bombyx mori* L, Mulberry varieties, Monsoon season, Rearing parameters

INTRODUCTION

Mulberry (*Morus* sp.) leaf is the only food and source of nutrition for the silkworm, *Bombyx mori* L. The growth and development of larva, and subsequent cocoon production are greatly influenced by nutritional quality of mulberry leaves (Krishnaswami, 1978). Narayanan *et al.*, 1996 and Siddhu *et al.*, 1969 have reported about the quality differences in leaves of different mulberry varieties. The quality of mulberry leaves differs significantly with factors such as soil fertility, agronomical practices,

Department of Applied Animal Sciences, Babasaheb Bhimrao Ambedkar. Central University, Vidya Vihar, Raibareli Road, Lucknow – 226025 (U.P.). E-mail: elango70@yahoo.com planting system and environmental conditions (Bongale *et al.*, 1991; Datta, 1992). Further, the quality of leaf is reported to vary with age and composition of leaf on the shoot. The nutritional value of top tender leaves was found to be higher as compared to middle and matured leaves or bottom leaves. Hence, the over matured leaves are not suitable for silkworm rearing. The aim of this study was to investigate the effect of different mulberry varieties on rearing performance of silkworm (*Bombyx mori*) in monsoon season of Uttar Pradesh.

MATERIALS AND METHODS

The study was carried out between 12 September to 10 October using CSR2 x CSR4 by feeding four different mulberry varieties namely AR-10, AR-12, AR14 and Chak majra (control) which are available in the experimental site of Babasaheb Bhimrao Ambedkar University, Lucknow. Standard rearing practices were used to assess the rearing performance which include larval weight, larval duration, effective rate of rearing, single cocoon weight, single shell weight, shell ratio percentage etc. Worms were fed four times in a day and four replicates were maintained for each mulberry variety. To minimize crowding at the beginning of 3rd instars, sum of 100 larvae were maintained in each replication which continued for further rearing. Equal quantity of leaves was offered to all the replicates since brushing until maturity. In addition, morphological parameters of worms were recorded.

RESULTS

The hybrid race CSR2 x CSR4 showed a differential consumption on four different varieties of mulberry leaves which influenced their growth. The mean weight of the larvae which fed leaves of AR-10 slightly larger (6.07 ± 0.31 g) than the worms fed AR-12 (5.13 ± 0.42 g), AR-14 (5.20 ± 0.20 g) and Chak majra (5.87 ± 0.64 g). However, the differential intake of larvae on four varieties of leaves did not differ significantly (ANOVA: $F_{3,}$ 11 = 3.67, *P* 0.062). The larvae which fed AR-12 showed highest

Table 8.1: Effect of different mulberry varieties on rearing performance of *B. mori*.

Parameters	Chak majra (Mean ± SD)	AR-10 (Mean ± SD)	AR-12 (Mean ± SD)	AR-14 (Mean ± SD)
Weight of matured larva	5.87 ± 0.64	6.07 ± 0.31	5.13 ± 0.42	5.20 ± 0.20
Growth rate (g/day)	0.218 ± 0.153	0.209 ± 0.148	0.220 ± 0.117	0.211 ± 0.123
Effective rate of rearing	91.67 ± 1.53	82.00 ± 4.36	68.00 ± 2.65	72.33 ± 2.52
Single cocoon weight (g)	1.30 ± 0.08	1.43 ± 0.04	1.41 ± 0.18	1.38 ± 0.14
Single shell weight (g)	0.29 ± 0.06	0.30 ± 0.08	0.29 ± 0.06	0.29 ± 0.01
Shell Ratio (%)	22.30 ± 3.38	20.97 ± 5.15	20.56 ± 1.74	21.01 ± 2.40
Larval duration (days)	28	28	28	28

growth rate 0.220 ± 0.117 g/day than the rest (Table 8.1). The effective rate of rearing was varied across the larval group which fed four different varieties of leaves, however the variety Chak majra fed larvae showed highest ERR 91.67 ± 1.53 (Table 8.1). The larvae which fed leaves of AR-10 showed highest single cocoon weight 1.43 ± 0.04 g and shell weight 0.30 ± 0.08 g. The shell ratio was ranged from 20.56 - 22.30% (Table 1). Nevertheless, there was no difference in the larval duration among four groups.

DISCUSSION

The present investigation revealed that rearing performance of silkworm was significantly influenced by the effect of different mulberry varieties. The rearing performance of current study suggests the suitability of variety AR-10 for subtropical climatic conditions. The current study substantiates the earlier reports that quality of mulberry leaves depend upon the variety (Toshio and Arai, 1963; Radha *et al.*, 1978). Tayade and Jawale, (1984) reported that the economic characters of hybrid silkworm cocoons are improved by feeding S-54 variety over K-2, Kosen and LM-2 varieties. Takashi (1961) observed better rearing performance by feeding Kanva-2 leaves over a local variety during late stage rearing. The current study substantiates the earlier reports that rearing performance of silkworm depends upon quality of mulberry varieties.

ACKNOWLEDGEMENTS

We thank Dr. Kamal Jaiswal, Department of Applied Animal Sciences, for providing laboratory facilities.

REFERENCES

Bongale, U.D., Chaluvachari & Rao, N.B.V. (1991). Mulberry leaf quality evolution and its importance. *Indian silk*, .29(11): 51 – 53.

Datta, R.K. (1992). *Guidelines for Bivoltine Rearing*. Central Silk Board, Bangalore, India, p. 18.

Krishnaswami, S. (1978). New technology of silkworm rearing. Bulletin Sericulture no.2 CSRTI, Mysore.Liaw, G. J. (1991) Effectiveness of artificial diets prepared from different varieties and maturity of mulberry leaves on development of silkworm *Bombyx mori* L. Chines. *J. Entomol.* 11(3): 260 – 263.

Narayanan, E.S., Kasiviswanathan, K. & Iyenger, S.M.N. (1996). Effect of varietals feeding, irrigation levels and nitrogen fertilization on the larval development and cocoon characters of *Bombyx mori* L. *Indian J. Sericulture,* 1(1): 13-17.

Radha, N.V., Lethchoumanane, S., Rajeswari & Obliswami, G. (1978). Effect of feeding with the leaves of different mulberry varieties on the races of silkworm. *All India symposiums on Sericulture Sciences UAS,* Bangalore (Abstract): 52

Sidhu, N.S., Kasiviswanathan, K. & Iyenger, S.M.N. (1969). Effect of feeding leaves grown under N.P and K. fertilization on the larval development and cocoon characters of *Bombyx mori* L. *Indian J. Sericulture,* 8(1): 55-60.

Takahashi, K., (1961) Report submitted to the Central Silk Board, Bombay, India.

Tayade, D.S. & Jawale, M.D. (1984) Studies on the comparative performance of silkworm races against different varieties of mulberry under Marathawada conditions. *Sericologia,* 24(3): 361-364.

Ito, T. & Arai, N. (1963). Food values of mulberry leaves for the silkworm Bombyx mori L. determined by means of artificial diets-Relationship between Kinds of mulberry leaves and larval growth. *Bulletin of Sericulture Expt.*, 18(4): 226-229.

9

BREEDING OF SPRING SPECIFIC BIVOLTINE SILKWORM HYBRID SBGP5 X SBGP22

Mir Nisar Ahmad, M.A. Khan, S.M. Quadir and Abad A. Siddiqui*

ABSTRACT

Although, the production of silk in India has increased considerably since last few decades, yet the country is not in a position to produce adequate bivoltine silk of 2A-4A grade to meet the domestic and international demand. In this direction, a breeding programme was initiated at Central Sericultural Research & Training Institute, Pampore in the year 1997 involving Chinese breeding resource material. The isolation of lines were initiated from F4 generation. The isolated lines were reared upto F13 generation and four lines viz. SBGP5, SBGP20, SBGP21, SBGP22 were finally selected on the basis of their performance. These four lines along with other races viz., CSR2, CSR5, SH6, NB4D2 and JP1B were utilized for hybrid evaluation. Thirty two hybrid combinations along with three control hybrids viz. SH6 x NB4D2, CSR2 x CSR4 and CSR2 x CSR5 were evaluated for three spring seasons during 2003, 2004 and 2005. The pooled data was evaluated by Evaluation Index Method of Mano *et al.*, (1993) and SBGP5 x SBGP22 was identified as one of the promising spring specific hybrid for commercial exploitation. In new hybrid, shell ratio is > 22% with pupation rate more than 90%. In the present paper, method of breeding of new lines has been discussed.

Key words: Silkworm, Isolation, Hybrid evaluation, Breeding

Central Sericultural Research & Training Institute (C.S.R.&T.I.), Pampore, Kashmir-192121 (J&K).

E-mail: mirnisarahmad@gmail.com

* P4, Basic Seed Farm, Central Silk Board, Manasbal, Kashmir (J&K)

E-mails: mirnisarahmad@gmail.com
csrtipper@vsnl.com
siddiquibad@yahoo.w.co.in

INTRODUCTION

Most of the bivoltine silkworm races used at commercial level are either exotic or evolved through hybridization. India had obtained numerous bivoltine hybrids periodically from Japan, Italy, France, China and Russia. In India, the concept of introducing hybrids for commercial silkworm rearing started during 1922 and Karnataka was the first in introducing crosses between pure Mysore and Japanese races, namely PM x C – Nichi and PM x S 6 (Nanavathy,1965), and thus the tradition of rearing hybrids instead of pure races came to being in south India (Datta, 1984). Realizing the better performance of hybrids over the pure lines such as fast and uniform growth, higher cocoon yield, crop stability; silkworm breeders started the development of new strains in India from sixties (Ahsan and Rao,1997). Harada (Anonymous, 1955) was the first to start the silkworm breeding in India and as a result, Kalimpong breeds were developed at RSRS, Kalimpong. Later on, NB series, CC1, CA2 etc., at CSR&TI, Mysore; SKUAST races at Kashmir ; PAM bivoltine races at CSR&TI, Pampore;SH6, YS3,SF19 and JD6 at RSRS, Dehradun likewise other breeds were developed at different stations (Trag *et al.*, 1992; Thangavelu,1997; Datta, 2000). The average cocoon yield of hybrids of races developed until early nineties, in India was about 35-40 kg/100dfls. Further, these bivoltine breeds were not so superior in their reeling characters to compete in international market. In this direction, a breeding programme was initiated at Central Sericultural Research & Training Institute, Pampore, in the year 1997 involving Chinese breeding resource material by adopting cross breeding and line separation method to develop new bivoltine breeding lines that have shell above 20% and suitable for rearing under temperate climatic condition.

MATERIALS AND METHODS

For the present study six single crosses of Chinese breeding material were utilized to evolve new bivoltine breeds. During early generations (F1-F3) of breeding (Spring 1997 to Autumn 1997) mass method of breeding with random mating (full-sib mating) was adopted for obtaining more gene recombinants and for increasing variation in breeding populations.

From F4 to F12 generations (Spring 1998 to Autumn 2000) each breeding line was reared in six cellular batches and pedigree/ line separation method of breeding was resorted for fixation of different lines. In each generation, batches in which shell ratio was above 20% and pupation rate was above 90% were selected for furthering generations. Selection was imposed both at batch and individual level as per the methodology advocated by JICA experts. To improve the cocoons characters i.e. single cocoon weight, single shell weight and shell ratio, single cocoon assessment was resorted during mid-generation of breeding and the cocoons which were above the batch average were selected as seed cocoons for the next breeding generation. Directional selection was also imposed on post cocoon parameters and the breeding lines/populations in which the filament characters were towards lower side such as filament length below 750 mts, reelability below 70%, neatness below 80 point and denier above 2.8 were culled during selection. The lines that have not responded to selection pressure and not exhibited the improvement in traits were also culled during breeding. In each generation data was collected on fecundity, hatching%, larval period, yield/10,000 larvae, single cocoon wt., shell wt. and shell ratio, pupation rate and post cocoon parameters viz., avg. single filament length, denier, reelability, raw silk% and Neatness. The standard rearing methods suggested by Krishnaswamy (1978) and Datta *et al.*, (1996) was adopted. After F12 generation four lines have shown uniformity in larval marking and cocoon shapes and stability in metric traits. The four selected lines viz., SBGP5, SBGP20, SBGP21 and SBGP22 were further reared for F13 and F14 generations during spring 2001 and spring 2002 respectively for study of the uniformity in desired traits. These four isolated lines along with other races viz., CSR2, CSR4, CSR5, SH6, and

NB4D2 were subjected to hybrid testing and 35 hybrids were evaluated in spring 2003 and spring 2004. Such type of study has also been reported by Mir Nisar *et. al.*, (2005). Mano's evaluation index (Mano *et al.*, 1993) was employed for ranking the hybrids. Denier studies were also conducted on the top ranking hybrids and the control hybrid.

Field trials of selected hybrids were also conducted at farmer's level under Research extension centers of CSR&TI, Pampore for two years (spring 2005 and 2006). In field trials 5 ounces of new hybrid along with the control hybrid were tested in Kashmir valley.

RESULTS AND DISCUSSION

Four new bivoltine silkworm lines viz., SBGP5, SBGP20, SBGP21 and SBGP22 were the final out come of the breeding project. Characteristics of new genotypes are summarized in Table 9.1. The fecundity in all the four lines is more than 500 with hatching more than 95%. All the four lines have the larval period of 26 days. SBGP5 yields 16.45 Kg cocoons per10000 larvae. The highest pupation rate of 96% was recorded in SBGP5 and the lowest 94.30 in SBGP21. The shell in all the isolated lines is above 21% with highest shell of 22.89% in SBGP5. The highest filament length of 970 meters was recorded in SBGP22. All these lines have thin denier recorded in the range of 2.40 to 2.04. These lines recorded above 75% reelability and above 85 points neatness.

Average rearing data of three top ranking hybrids viz. SBGP5 × SBGP22, SBGP22 × SBGP5 and CSR2 × SBGP22 along with control hybrid SH6 × NB4D2 is summarized in Table 9.2. In the new hybrids larval period is 12 hours lesser than the control hybrid SH6 × NB4D2. The highest shell of 42 cg was recorded in SBGP5 × SBGP22 with recording higher shell ratio (22.44%), higher filament length of 1005 meters, lesser denier (2.25) and higher reelability (77%). All the three new hybrids recorded 85 points neatness. Denier studies were also conducted in the three new hybrids and the control hybrid. Denier was recorded in the cocoon

Table 9.1: Mean values of different traits of new SBGP Lines

Traits	SBGP5	SBGP20	SBGP21	SBGP22
Larval marking	Plain	Plain	Plain	Plain
Cocoon shape	Oval	Constricted	Constricted	Constricted
Fecundity	646	523	583	549
Hatching%	96.61	96.48	94.30	95.85
Larval period(D:H)	26:00	26:00	26:00	26:00
Yield/10000 larvae(kg)	16.45	15.24	15.65	16.29
Pupation rate	96.0	95.0	94.0	95.0
Single cocoon wt.(g)	1.66	1.75	1.74	1.83
Single shell wt(g)	38.0	38.0	39.0	40.0
Shell ratio%	22.89	21.61	22.41	21.85
Av. Filament length (mts)	920	875	925	970
Denier	2.10	2.25	2.40	2.04
Reelability%	75	80	80	75
Neatness	85	85	80	87

Table 9.2: Mean values of different traits in new hybrids

Hybrid	Hatching %	Larval Period D:H	Pupation rate	Cocoon yield /10,000 larvae (kg)	Single cocoon wt(g)	Single shell wt (cg)	Shell ratio %	Av. Filament length (mts.)	Denier	Reelability %	Neatness
SBGP5 X SBGP22	97.75	25:12	95.8	18.76	1.87	42.0	22.44	1005	2.25	77	85
SBGP22 x SBGP5	97.48	25:12	93.3	17.39	1.78	39.0	21.83	991	2.30	75	85
CSR2 x SBGP22	97.25	25:12	87.8	17.46	1.74	39.0	22.16	950	2.35	75	85
SH6 x NB4D2 (Control)	97.82	26:00	96.7	14.20	1.80	34.0	18.88	850	3.20	70	80

Table 9.3: Field performance of SBGP5 × SBGP22 and control SH6 × NB4D2 (mean of two rearings) reared during spring 2005 and 2006.

Hybrids	Yield/100dfls (kg)	Single cocoon wt.(g)	Single shell wt.(cg)	Shell ratio	Av. Filament length(m)	Denier
SBGP5XSBGP22	52.20	2.02	43.0	21.28	950	2.4
SH6XNB4D2 (control)	45.50	2.15	38.0	17.67	825	3.2
Percent gain over control	14.72	-6.04	13.15	6.81	15.15	-25.00

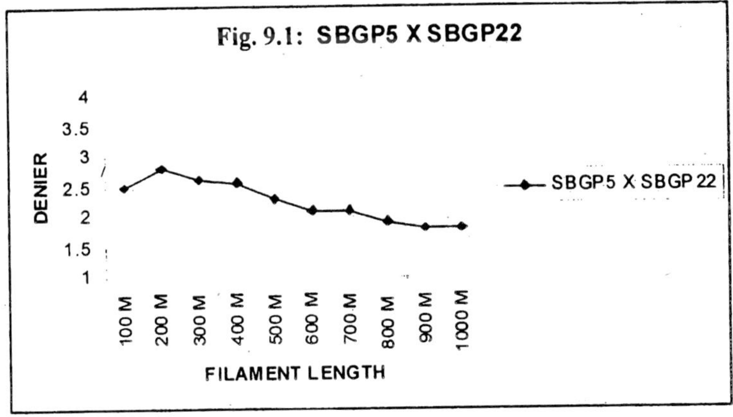
Fig. 9.1: SBGP5 X SBGP22
DENIER
4
3.5
3
2.5
2
1.5
1
100 M
200 M
300 M
400 M
500 M
600 M
700 M
800 M
900 M
1000 M
FILAMENT LENGTH
SBGP5 X SBGP 22

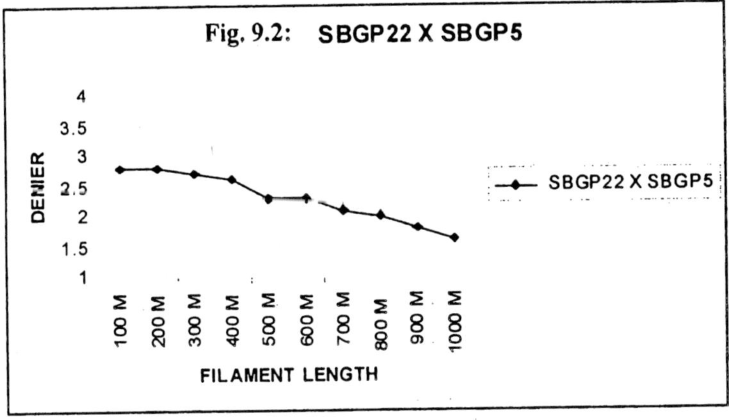
Fig. 9.2: SBGP22 X SBGP5
DENIER
4
3.5
3
2.5
2
1.5
1
100 M
200 M
300 M
400 M
500 M
600 M
700 M
800 M
900 M
1000 M
FILAMENT LENGTH
SBGP22 X SBGP5

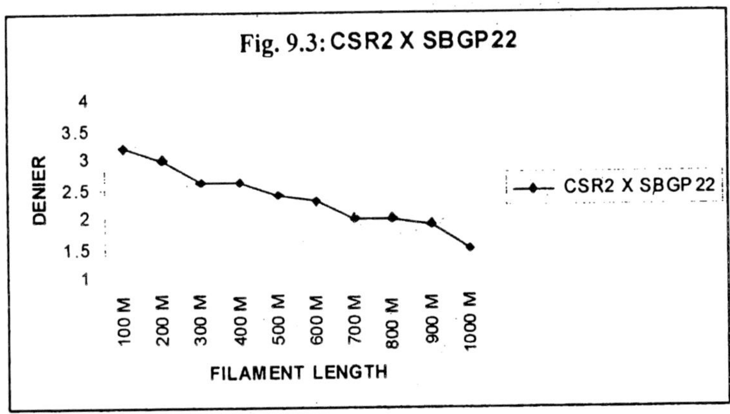
Fig. 9.3: CSR2 X SBGP22
DENIER
4
3.5
3
2.5
2
1.5
1
100 M
200 M
300 M
400 M
500 M
600 M
700 M
800 M
900 M
1000 M
FILAMENT LENGTH
CSR2 X SBGP 22

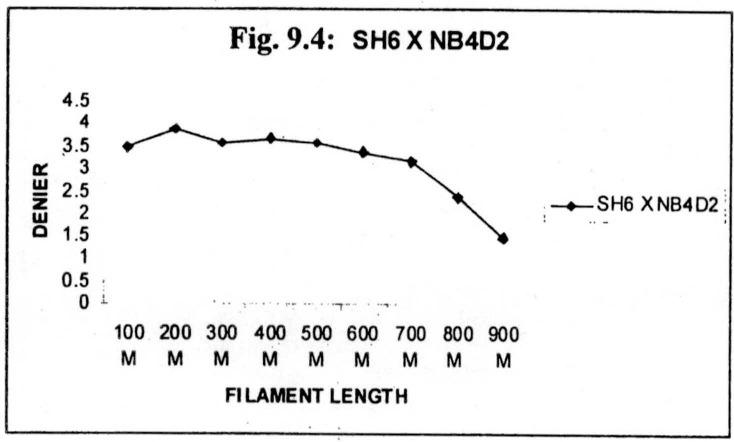

filament from outside to the inside of the cocoon. The study reveals lesser deviation in cocoon filament of the new hybrids as compared to SH6 × NB4D2. The data is represented in Figs. 9.1-9.4.

SBGP5 × SBGP22 was tested in field along with control hybrid SH6 × NB4D2 in three sericulturally important districts of Kashmir valley viz., Anantnag, Baramulla and Pulwama for two spring seasons during years 2005 and 2006. Perusal of the field data reveals the superiority of the new hybrid SBGP5 × SBGP22 over the conventional hybrid SH6 × NBD2. The new hybrid recoded a gain of 14.72%, 13.15%, 6.81% and 15.15% in cocoon yield, shell weight, shell ratio and average filament length respectively, over the control hybrid.

For the development of new genotypes, proper selection of desirable traits is to be done and it is possible when the heterozygous breeding populations are available. If variation is less in any breeding population then it is difficult to select desirable individual (cocoons) and improvement in different traits is difficult to achieve (Siddiqui *et al.*, 2005). Therefore, to obtain heterozygous population as well as more gene recombinants breeding batches, in the present study, during earlier generation of breeding (F1-F3) were reared in mass coupled with random mating. For effective selection, it is advocated by different breeders that breeding populations (families) should be reared at least in 5-8 cellular batches in mid generations of breeding. In the present study also each breeding line was reared in six cellular batches between F4 and F12 generations and out of that only one or two batches that have the pupation rate, cocoon wt, shell weight and shell ratio above the average of six batches were only selected. Selection was made both at batch and individual level. According to Kobari and Fujimoto (1966), there is co-relation between different traits and selection for one character affects the other traits of economic importance. The shell weight and the shell ratio have the negative co-relation with survival percentage, therefore, the selection was made both at batch (for selection on survival) and individual level (for selection on cocoon traits) to maintain a balance between them. Utilization of hybrid vigour for the development of new genotypes was introduced in 1901 by Toyama (1906) and with this strategy in present study different races were crossed to utilize their hybrid vigour & variation as a tool to develop new genotypes. Simultaneously, in India, utilization of exotic hybrids has also resulted in the evolution of a series of productive bivoltine CSR breeds (Basavaraja *et.al.*, 2003).

It is evident from the lab as well as the field data that the new hybrids have a great potential for commercial exploitation and based on the performance in various rearing and reeling parameters SBGP5 X SBGP 22 is recommended for commercial exploitation during spring season, the main raring season in North India.

REFERENCES

Ahsan, M.M. & Mohan, R. (1997). Progress through silkworm breeding in India. Base paper, *Current Technology Seminar*, 18- 19 Sept, CSR&TI, Mysore, India.

Annonymous (1955). *Annual report. CSR&TI*, Behrampore, India.

Basavaraja, H.K., Kumar, S.N., Kariappa, B.K. & Dandin, S.B. (2003). Constraints, present status and prospects of silkworm breeding, Base paper, *Silkworm breeders summit*, 18-19 July, 2003, APSSRDI, Hindupur, India.

Datta, R. K. (1984). Improvement of silk worm races (*Bombyx mori* L.) in India. *Sericologia*, 24: 393 – 415.

Datta, R.K., Basavaraja, H.K. & Mano, Y. (1996). Manual on bivoltine rearing race maintenance and multiplication, Published by CSR&TI, Mysore, India.

Kobari, K. & Fujimoto, N. (1966). Studies on the selection of cocoon filament length and cocoon filament size in *Bombyx mori* L. Nissenzatsu, 35(6): 427-434.

Krishnaswamy, S. (1978). New technology of silkworm rearing, CSR&TI, Bull no. 2, CSRTI, Mysore, India.

Mano, Y., Kumar, N.S., Basavaraj, H.K., Malareddi, N. & Datta, R. K. (1993). A new method to select promising silkworm breeds combinations. *Indian Silk*, 31: 53.

Ahmad, M.N., Khan, M.A. & Quadir, S.M. (2005). Evaluation for identification of spring specific silkworm *Bombyx mori* L. hybrids for commercial exploitation under Kashmir climatic conditions. *Proceedings Volume: II of the XX Congress of the International Sericultural Commission France*, held at Bangalore India on 15-18 December. pp. 351-356.

Nanavathy, M.M. (1965). Silk from Grub to Glamour. Paramount Publishing House, Bombay, p. 277

Siddiqui, A.A., Chakrabarti, S., Lochan, R. & Chauhan, T.P.S. (2005). Development of new bivoltine silkworm, DUN breeds, by cross breeding method. *Casp. J. Env. Sci.*, 3: 163-168.

Trag, A.R., Kamili, A.S., Malik, G..N. & Kukiloo, F. A. (1992). Evolution of high yielding bivoltine silkworm (*Bombyx mori* L.), genotypes, *Sericologia*, 32(2): 321-324.

Thangavelu, K. (1997) Silkworm breeding in India and over view. *Indian Silk*, June, 5-13.

Toyama, K (1906). Breeding methods of silkworm. *Sangyo Shimpo*, 158: 282-286.

EVALUATION OF REARING AND SILK PARAMETERS IN ELITE BIVOLTINE SILKWORM GERMPLASM

M. Muthulakshmi, N. Balachandran, B. Mohan, S.A. Hiremath, P.R. Koundinya, R.K. Sinha and S.M.H. Qadri

ABSTRACT

Central Sericultural Germplasm Resources Centre, Hosur, is the largest gene bank in India for silkworm genetic resources, which maintains 432 *Bombyx mori* germplasm accessions representing 16 countries of the world. About 79% of the total collection (339 accessions) comprises bivoltine genetic resources, which are considered important recently in breeding point of view for improving the silk production and to obtain international grade of silk quality. CSGRC, Hosur, is conserving these genetic resources after standardized characterisation and evaluation protocols and publishing the information for the use of breeders to select the desired genotype for their breeding plan. In this line, 209 bivoltine silkworm germplasm were evaluated for rearing and silk parameters over the data generated for the period of 4 years using Mano's cumulative evaluation indices. Elite bivoltine germplasm viz., BBI-0255, BBE-0224,and BBE-0268 were identified for rearing, silk quantity and quality parameters with multiple traits performing better than the ruling popular bivoltine silkworm germplasm CSR-2 and CSR-4.

Key words: *Bombyx mori*, Germplasm, Silkworm, Evaluation Index.

INTRODUCTION

Genetic resources contribute as raw materials for crop improvement. The success of breeding depends on the initial

Central Sericultural Germplasm Resources Centre, Thally Road, Hosur-635109, Tamil Nadu, India.

E-mails: lahshuricsgrc@yahoo.co.in; balucsgrc@yahoo.co.in

selection of parents, their effective utilization in desirable combinations. It also depends on the ability of breeds to assemble and recombine the genetic variability, to extract the potential gene combinations from the gene pool based on phenotypic expression leading to genetic fixation of the traits over generation (Nirmal Kumar and Sreeramareddy, 1994). In the absence of proper conservation system old traditional races like Chotopolu and ancient breeds like Ichot, Nismo and Itan are extinct (Thangavelu, *et al.*, 2000). Central Sericultural Germplasm Resources Centre (CSGRC), Hosur is maintaining 359 bivoltine silkworm germplasm accessions, which includes 20 mutant genetic stocks. These accessions represent 13 countries including India. (Table 10.1). The collection in gene bank includes old popular breeds and also newly evolved breeds. The assemblage of germplasm does not end with regular collection, characterisation, evaluation and conservation, but requires presentation of the important data analysed in various ways by identifying the elite germplasm superior for one or more traits to promote their utilisation by the breeder in the crop improvement programme. Evaluation is the ultimate objective in germplasm management to assess the fitness of germplasm for field use, based on the data generated on various traits (Thangavelu, *et al.*,1997). Preliminary evaluation was done for 22 growth and reproductive traits and 16 reeling traits. Silkworm germplasm is mainly utilized for quality improvement and commercial exploitation. Screening out the promising pure breeds from the germplasm stocks is an important duty of the curator of gene bank maintenance.

The genetic improvement in the evolved breeds are still required, since the genetic distances among the Indian breeds are very narrow because some of the breeds are repeatedly used in the breeding programme. Further, it is also known that some of the genetic characters are linked and hence even if the breeder aims at improving one/more-desired character, some other desired character gets altered. (Dalton, 1987; Falconer, 1989; Anon, 1997.)

It is important that the genetic difference between parents should be wide to get heterosis and either positive or negative heterosis is expressed in the cross (Dalton, 1987). Varied silkworm germplasm stocks contribute immensely to the development of viable and hardy silkworm breeds for commercial exploitation. (Nirmalkumar and Sreerama Reddy, 1994).The very purpose of the existence of a conservation centre is to collect, characterize, evaluate and conserve the germplasm material for which it is meant for and also to prevent these valuable genetic material evolved very long period from extinction. Not only such conservation centers perform the doing the job of curator, but also properly evaluate and document the data on different characterisation and evaluation in proper and easily recoverable format in the form of databases. Such well characterized and evaluated silkworm germplasm are very handy tools. Breeders want such proven and thoroughly evaluated breeding materials for inclusion as breeding lines in the breeding programmes in addition to evaluation different region and season specific breeds for different agroclimatic conditions. In such context, this study would be of much value. Hence, an attempt has been made to identify suitable bivoltine silkworm germplasm with special characteristics to meet the present demand for utilizing them in various silkworm breeding programmes.

MATERIALS AND METHODS

At present 359 bivoltine accessions are being maintained at CSGRC, Hosur. Out of which, 209 accessions that completed reeling analysis were considered for the present study. Bivoltine accessions were reared once in a year and the average of four crops from 2003 to 2006 has been taken for assessment. DFLs of these accessions are preserved in cold storage following 10 months preservation schedule and are brushed as composite population to avoid inbreeding depression as well as genetic erosion and to maintain the gene pool as far as possible. 40 disease free layings are taken at random in each accession and divided into two batches of 20 dfls each. The composite layings are prepared only

Table 10.1: Country wise collections of Bivoltine silkworm Genetic resources at CSGRC, Hosur

Country name	*Indigenous*	*Exotic*	*Total*
Brazil	-	3	3
China	-	40	40
France	-	11	11
India	159	28	187
Indonesia	-	1	1
Iraq	-	1	1
Japan	-	64	64
Poland	-	3	3
Russia	-	19	19
South Korea	-	1	1
Thailand	-	4	4
Ukraine	-	2	2
Vietnam	-	3	3
Total	159	180	339

after the body pigmentation by taking approximately 50 eggs from each laying. All the pieces from 20 laying sources are pasted on a slightly thick brown paper and wrapped in white fine tissue paper after drying. Thus, each composite laying consists of about 1000 individual eggs pooled from 20 dfls. Similarly one more composite laying is prepared from the rest of the 20 layings. Thus, in each accession two composite layings are prepared for brushing as two replications. The replications are maintained for data collection and analysis of economically important quantitative and qualitative characters. Rearing was conducted by following standard rearing procedures (Krishnaswami, 1978). These genetic resources were evaluated for 22 growth and reproductive parameters and 16 reeling and quality parameters.

For the present analysis the data generated on the following important traits were taken into consideration viz; fecundity (No.), weight of 10 larvae (g),total larval duration (hrs) cocoon yield/ 10000 larvae by no, cocoon yield/10,000 larvae by wt. (kg), pupation percentage, shell ratio (%), cocoon yield /100 dfls (kg), filament length (mts), denier, renditta (kg), neatness (%), raw silk recovery (%). The data on these parameters for four years from 2003-2006 were analyzed based on query in indigenously developed database Silkworm Germplasm Information System (SGIS). The query on special characters was made in comparison with the two control accessions CSR-2 and CSR-4.

In order to judge the superiority of silkworm varieties impartially a common index has been used. To accomplish the objective an Evaluation Index (EI) developed (Mano, 1993) was utilized. The EI is an index of multiple traits or as a performance index, which is a single valued measure of the multiple trait performance of a population. EI was obtained as follows

$$\text{E.I.} = \frac{A - B}{C} \times 10 + 50$$

Where A = mean of the particular trait

B = overall mean of the particular trait

C = standard deviation

10= fixed value

50 = constant

In silkworms a large number of hybrids are tested and promising ones are selected based on the economic traits. EI is the strategy, which can increase the precision of the selection of hybrids among the array of breeds and giving adequate weightage to all the yield component traits. As such, it will be fair and precise to take the decision based on the multiple traits spanning the entire growth period, so that adequate weightage is given to various characters that are manifested in the entire growth phase of the silkworm.

RESULTS AND DISCUSSION

In general, silkworm germplasm is basis for quality improvement and commercial exploitation. Screening the promising pure breeds from the germplasm stocks is an important duty of the persons responsible for gene bank maintenance (Kumaresan, 2004). Accessions with special characters for rearing as well as reeling were selected based on better performance over four years. 209 bivoltine accessions have been evaluated for the major economically important eight rearing traits and five reeling traits (Table 10.2). These accessions are compared with CSR-2 and CSR-4, which are the ruling bivoltine pure breeds utilized in the bivoltine hybrid seed production at commercial level at present.

Higher fecundity (more than 580 eggs/dfl) was recorded in 19 potential accessions, which is 50 and 30 eggs more than that of the average fecundity of CSR-2 and CSR-4 respectively. In the case of 10 larval weight more than 42 gms was recorded in 16 accessions with maximum of 44 gms in accession no. BBE-35, which is 4-6 gms higher than CSR-2 and CSR-4. Shorter larval duration of 22 days was observed in 14 accessions, which is 2 days lesser than ruling breed, where the larval period is 24 days. Higher pupation rate (94-96%) was noticed in 21 accessions whereas, in CSR-2 and CSR-4 it is 92%. Higher cocoon yield per 10000 larvae by numbers was recorded in 26 accessions ie, 9750- 9870, while it was 9519 in CSR-2 and 9376 in CSR-4. Cocoon yield per10000 larvae by weight was ranged between 21-19 kg in 27 accessions, and it was 19.9Kgs in CSR-2 and 17.2 Kgs in CSR-4. Cocoon yield/100 dfls ranged from 85-77 kg/ 100dfls in 18 accessions, but in CSR-2 it was 79.6 Kgs. and 68.7 Kgs in CSR-4. Under the important reeling parameters analysed high silk ratio of 24-22% was recorded in 22 accessions, where as it was 21.6 and 22.1 in CSR-2 and CSR-4 respectively. Higher total filament length of more than 1000 meters was recorded in 14 accessions (1190-1000) with 1190m in accession BBI-0275,

Table 10.2: Comparative performance of elite bivoltine germplasm with ruling bivoltine pure breeds

Character	Value	Range	No. of Accessions	List of Accessions	CSR-2	CSR-4
Higher Fecundity (No.)	>580	580 - 632	19	BBI-0277, BBE-0179, BBI-0255, BBE-0177, BBI-0046, BBE-0225, BBE-0013, BBE-0272, BBE-0186, BBE-0216, BBE-0201, BBE-0224, BBE-0181, BBI-0286, BBE-0187, BBI-0172, BBE-0006, BBI-0276, BBE-0010	536	559
Higher larval weight (g.)	>42	42 - 44	16	BBE-0035, BBE-0004, BBI-0129, BBI-0079, BBI-0113, BBE-0206, BBE-0329, BBE-0177, BBE-0202, BBI-0081, BBI-0137, BBE-0268, BBI-0066, BBI-0045, BBI-0091, BBE-0212	37.4	38.9
Shorter Larval duration (Days)	<23	23 - 22	11	BBE-0021, BBI-0073, BBI-0072, BBE-0022, BBI-0069, BBI-0103, BBE-0024, BBE-0020, BBE-0014, BBE-0016, BBE-0015	24	24
Higher Pupation rate (%)	>94	94 - 96	21	BBE-0194, BBI-0058, BBE-0210, BBI-0068, BBE-0212, BBE-0037, BBI-0255, BBI-0207, BBI-0061, BBI-0071, BBI-0077, BBE-0005, BBE-0015, BBI-0044, BBE-0222, BBI-0103, BBI-0257, BBE-0042, BBE-0196, BBI-0063, BBI-0069	92.6	92.03
Higher cocoon yield/10000 larvae (No.)	>9750	9750 - 9870	26	BBI-0106, BBI-0117, BBE-0194, BBE-0222, BBE-0196, BBI-0097, BBE-0199, BBE-0193, BBI-0103, BBI-0137, BBI-0113, BBI-0132, BBI-0207, BBI-0123, BBE-0224, BBI-0091, BBE-0005, BBI-0239, BBI-0124, BBE-0226, BBI-0136, BBE-0001, BBI-0140, BBE-0260, BBI-0098, BBI-0255	9519	9376
Higher cocoon yield/10000 larvae (Kg.)	>19	19 - 21	27	BBI-0113, BBE-0010, BBI-0133, BBE-0268, BBE-0004, BBI-0044, BBI-0255, BBE-0035, BBE-0006, BBI-0134, BBI-0129, BBI-0137, BBE-0224, BBI-0126, BBI-0091, BBI-0086, BBI-0068, BBI-0056, BBE-0003, BBE-0225, BBI-0172, BBI-0079, BBE-0050, BBE-0043, BBI-0127, BBE-0270, BBI-0045	19.9	17.2
Higher cocoon yield/100dfls (Kg.)	>77	77 - 85	18	BBI-0113, BBE-0010, BBI-0133, BBE-0268, BBE-0004, BBI-0044, BBI-0255, BBE-0035, BBE-0006, BBI-0134, BBI-0129, BBI-0137, BBE-0224, BBI-0126, BBI-0091, BBI-0086, BBI-0056, BBI-0068	79.6	68.7
Higher Cocoon Shell Ratio (%)	>22	22 - 24	22	BBE-0225, BBE-0226, BBE-0185, BBE-0182, BBI-0275, BBE-0269, BBE-0262, BBE-0266, BBE-0236, BBE-0272, BBE-0263, BBE-0186, BBE-0197, BBI-0325, BBE-0247, BBE-0193, BBE-0181, BBE-0267, BBE-0265, BBE-0224, BBI-0239, BBE-0270	21.6	22.1
Higher Filament length (m.)	>1000	1000 - 1190	14	BBI-0275, BBE-0267, BBE-0266, BBE-0265, BBI-0326, BBE-0225, BBE-0262, BBE-0263, BBE-0270, BBE-0268, BBE-0171, BBE-0177, BBE-0202, BBE-0214	1077	938
Low denier (d)	<2.1	2.1-1.61	21	BBE-0026, BBE-0195, BBI-0062, BBI-0126, BBI-0115, BBI-0109, BBE-0024, BBE-0250, BBE-0219, BBE-0025, BBE-0214, BBE-0260, BBE-0035, BBI-0085, BBE-0015, BBI-0073, BBE-0051, BBE-0018, BBE-0050, BBE-0041, BBI-0255.	2.69	2.6
Low Renditta (Kg.)	<6.0	6.0 - 4.8	14	BBE-0182, BBE-0262, BBI-0325, BBE-0263, BBE-0187, BBI-0326, BBE-0227, BBE-0212, BBE-0216, BBE-0188, BBE-0185, BBE-0202, BBE-0201, BBE-0266	6.56	7.48
Higher Raw silk recovery (%)	>78	78 - 90	21	BBE-0182, BBE-0262, BBE-0216, BBI-0235, BBE-0013, BBI-0325, BBE-0187, BBE-0266, BBI-0275, BBE-0263, BBE-0227, BBI-0172, BBE-0267, BBI-0083, BBE-0219, BBI-0054, BBI-0286, BBE-0214, BBI-0273, BBE-0212, BBE-0193	65.5	66.2
High Neatness (%)	>94	94 - 98	22	BBI-0124, BBE-0185, BBI-0106, BBI-0125, BBI-0132, BBE-0177, BBI-0112, BBI-0105, BBI-0108, BBE-0188, BBI-0134, BBE-0187, BBE-0265, BBI-0072, BBI-0285, BBI-0327, BBI-0127, BBE-0261, BBE-0268, BBE-0229, BBI-0294, BBI-0136	90.0	88.5

Table 10.3: List of top ranking bivoltine germplasm with multiple qualifying parameters

Ranking of Germplasm	*No. of Parameters Qualified*	*Character**
BBI-0255	6	1(619.5), 4(94.324), 5(9750.167), 6(19.917), 8(79.626), 10(2.1),
BBE-0224	5	1(591.167), 5(9776), 6(19.533), 7(22.284), 8(78.081)
BBE-0268	5	2(42.311), 6(20.033), 8(80.203), 9(1026), 13(94.5),
BBE-0187	4	1(586.333), 11(5.6), 12(80.85), 13(96),
BBE-0035	4	2(44.09), 6(19.712), 8(78.857), 10(2.055),
BBE-0262	4	7(23.089), 9(1038.4), 11(4.85), 12(90.115),
BBE-0266	4	7(23.06), 9(1085), 11(6), 12(80.85),
BBE-0177	4	1(610), 2(42.736), 9(1017.85), 13(97.5),
BBE-0263	4	7(22.715), 9(1037.65), 11(5.515), 12(80.65),
BBE-0212	4	2(42.054), 4(94.824), 11(5.895), 12(78.495),

*1.Fecundity (No.),2. Weight of 10 larvae (g.),3. Larval duration (Days),4. Pupation percentage,5. Cocoon yield/10000 larvae by no,6.Cocoon yield/10,000 Larvae by wt. (Kg),7. Cocoon yield /100 dfls (Kg.), 8. Cocoon shell ratio (%),9. Filament length (m),10.Denier(d),

11. Renditta (Kg.),12. Raw silk recovery (%),13. Neatness (%).

whereas CSR-2 and CSR-4 had lower filament length of 1077m and 938m respectively. In 21 accessions, low denier of less than 2.1(1.61-2.1) was recorded which are much superior when compared with CSR-2 (2.69) and CSR- 4(2.6). The renditta recorded was 6.56 in CSR-2 and 7.48 in CSR-4, whereas low renditta of less than 6.0(4.8-6.0) was recorded in 14 accessions. Raw silk recovery was in the range of 78-98% in 21 accessions, but in the case of the ruling breeds it was less than 70%(65.5% in CSR-2 & 66.2% in CSR-4). 22 accessions showed more than 94% for neatness, whereas CSR-2 and CSR-4, the values were 85% and 88.5% respectively.

Ranking of bivoltine silkworm Germplasm

To identify the silkworm germplasm accessions which has the potentiality to perform better for more than one trait, ranking analysis was done based on Mano's Evaluation Index for selected economically important eight rearing and five reeling traits (Table 10.3) and those accessions which ranked for more number of traits were listed out as the potential bivoltine silkworm germplasm accessions among the germplasm stocks maintained at CSGRC, Hosur. Three elite accessions, BBE-0224, BBE-0268 and BBI-0255 were identified for possessing better ranking for five and six specialized characters respectively. Seven accessions *viz*; BBE-187, BBE-0035, BBE-0262, BBE-0266, BBE-177, BBE-263, BBE-212 were found to possess four specialized traits.

Genetic resources are the important basic materials for continuous crop improvement either in breeding programmes aimed at higher productivity, developing specialized breeds or for inclusion in evaluation studies in different agroclimatic conditions, screening against abiotic and biotic stress environments. During last four decades, several bivoltine breeds were evolved (Krishnaswami, 1983; Datta, 1984; Govindan et al, 1996; Thangavelu, 1997). Since the multivoltine races are poor in productivity and cocoons produced from these races do not yield quality silk yarns suitable for power loom, bivoltine breeds are given much importance to compete with the international silk market. Heterosis breeding in silkworm has substantially contributed to the increase in cocoon production and in improving the quality of raw silk (Govindan et.al., 1996). It is obvious that the silkworm Germplasm constitute the potential raw material having wide variation in their genotypic expressions besides additive effect. The silkworm Germplasm available at CSGRC, Hosur are predominantly of evolved breeds in addition to some of the old races. Most of the evolved breeds were not tested in the fields due to lack of major information on genetic constituents and lack of diversity among the parents used. This study throws light on the availability of some of the potential bivoltine silkworm accessions, which are superior to the ruling silkworm pure breeds. Many of the characterization parameters based on morphological traits also showed very high degree of variability among the 209 bivoltine accessions studied.

REFERENCES

Anonymous (1997). *Principles and techniques of silkworm breeding.* ESCAP, Oxford & IBH Publishing Co. Pvt. Ltd., United Nations, New York.

Dalton, D.C. (1987). *An introduction to practical animal breeding.* Second edition, English Language Book Society/Collins.

Falconer, D.S. (1989). *Introduction to quantitative genetics.* Third Edition, English Language Book Society, Longman.

Govindan, R., Rangaiah, S., Narayanaswamy, T.K. & Devaiah, M.C. (1996). Genetic divergence among multivoltine genotypes of silkworm, *Bombyx mori* L. *Nissenzatsu,* 36: 427-434.

Krishnaswami, S. (1978). *New Technology of Silkworm Rearing.* Bulletin No.2, Central Sericultural Research and Training Institute, Central Silk Board, Mysore, India.

Krishnaswami, S. (1983). Evolution of bivoltine races for traditionally multivoltine areas of south India. *Indian Silk,* 22: 3-11.

Kumaresan, P., Sinha, R.K., Mohan, B. & Thangavelu, K. (2004). Conservation of multivoltine silkworm (*Bombyx mori* L.) germplasm in India-An overview. *Int. J. Indust. Entomol.,* 9:1-13.

Mano, Y., Kumar, N.S., Basavaraja, H.K., Reddy, N., Mal & Datta, R.K. (1993). A new method to select promising silkworm breeds/ combinations. *Indian Silk,* 31: 53.

Kumar, N.S. & Reddy, G..S. (1994). Evaluation and selection of potential parents for silkworm breeding. In: *Silkworm breeding.* (ed.Reddy, G.S.). Oxford &IBH Publishing Co. Pvt. Ltd., New Delhi, pp. 63-78.

Thangavelu, K.. Mukherjee, P., Sinha, R.K., Mahadevamurthy, T.S., Mukherjee, S., Sahni, N.K., Kumaresan, P., Rajarajan, P.A., Mohan, B. & Sekar, S. (1997). *Catalogue on Silkworm (Bombyx mori L.) Germplasm.* Vol.-1 (Pub.). Central Sericultural Germplasm Resources Centre , Hosur, p.138.

Thangavelu, K., Sinha, R.K., Mahadevamurthy, T.S., Radhakrishnan, S., Kumaresan, P., Mohan, B., Rayaradder, F.R. & Sekar, S. (2000). *Catalogue on Silkworm Bombyx mori L. Germplasm.* Vol.-2 (Pub.) Central Sericultural Germplasm Resources Centre, Hosur, p. 138.

11

BREEDING PHENOMENON OF SEX LIMITED SILKWORM GENOTYPES *BOMBYX MORI* L. ON COCOON COLOUR

*T.P.S. Chauhan and Mukesh Tayal**

ABSTRACT

Sex limited viable silkworm genotypes are of great importance to the silkworm egg production centres for preparation of hybrid silkworm seed with quality and at low cost of production. Many sex markers have been developed in silkworm *Bombyx mori* L., out of which sex specific larval marking and sex specific cocoon colour have been commercially exploited in sericulture industry for hybrid seed production.

Sex separation on cocoon colour has been found most appropriate and accurate in preparation of silkworm hybrid. However, sex limited silkworm races on cocoon colour have been reported to exhibit poor fecundity and poor survival. This makes them uneconomical to rear as parent race at P1 level and their use in the hybrid seed production.

The authors of present study used cross breeding technique to evolve sex limited silkworm genotypes with higher viability, high cocoon shell ratio and fecundity on par with their breeding parents. The cross breeding procedure, the generation response to economic rearing and cocoon characters have been described in the study. The response of selection (family and directional) to cocoon characters as against F1 generations and the mid parental values of the breeding parents have been analyzed.

Key words: Cross breeding, *Bombyx mori*, Sex Limited genotypes on cocoon colour.

Regional Sericultural Research Station, Miransahib, Jammu181101 (J&K).
*Regional Sericultural Research Station, Sahaspur, Dehradun-248001

INTRODUCTION

Sex linked characters in the insect were used in the past to determine the sex specific environmental changes in the species with the gradual estimate of physiological and biochemical parameters. Sexual dimorphism or sex marked characters in insects helped in studying the sex specific sensitivity to the environment and pathogens causing different diseases. Sex linked morphological characters in silkworm *Bombyx mori* L., brought a phenomenal change in the hybrid silkworm seed production in the commercial sericulture.

Ad initio sex specific larval marks on the ventral side of 11th and 12th abdominal body segment were used in identification of male and female larvae of silkworm *Bombyx mori* L. These sex specific identifications marks were found difficult to locate even in the late instar larvae. Later on identification of male and female were resorted at pupal stage using sex specific marks present on the ventral side of pupae. The female pupae have a longitudinal slit or a 'x' mark on the ventral side of the 8th segment and male has a dot mark between 8th and 9th segment on ventral side. Separation of male and female pupae in this method is comparatively easy but sex separation with this method requires huge skilled man power and bear risk of pupal injury due to handling. The whole process is too cumbersome and laborious with only 90-95% accuracy.

The sex- limited characters have been widely used in sericulture industry for the preparation of hybrid silkworm seed. The sex specific cocoon colour, though not very popular in other sericulturally advanced country can be used very effectively in the preparation of multi. x bi. hybrid seed in India saving considerable labour man days and revenue loss in the grainage from reeling cocoons of undesirable sex being cut for sex separation (Mundukur *et al.*, 2002). Under present study, sex specific cocoon colour silkworm genotypes have been evolved.

MATERIALS AND METHODS

Sex specific cocoon colour silkworm races of *Bombyx mori* L. were evolved through cross breeding between productive silkworm genotypes and donor sex limited genotypes on cocoon colour. Donor parent sex limited silkworm races CC1 SL and NB4D2 SL were obtained from CSGRC, Central Silk Board, Hosur, T.N., India. Other productive bivoltine parents were drawn from the silkworm germ plasm of the station. Productive silkworm races namely O3 and D2 were crossed with NB4D2 SL and CC1 SL donor races. Total four F1 hybrids viz., NB4D2 SL x O3, NB4D2 SL x D2, CC1 SL x O3 and CC1 SL x D2 were prepared and reared in mass of 5 DFLs in laboratory under controlled conditions. Mass rearing and random mating were followed till F2 generation. Lines were drawn after F2 generation selecting cocoons with different shapes and colours. Selected cocoons were individually assessed for shell quality and cocoons with higher, medium and lower shell ratio were grouped separately for breeding. Each group or family with specific cocoon characters was termed as line.

Lines were reared in cellular batches on quality mulberry leaf following standard rearing technique. Cocoon selection on target characters was resorted to after every generation rearing. Individual cocoon assessment was carried out from F3 to F7 generation and there after mass cocoon assessment of 50 males and 50 females was taken up till the completion of the study. Cocoon selection on cocoon shell index values as suggested by Yamamoto (1989) was carried out from F3 to F7 generation.

$$\text{Cocoon shell index} = \frac{\text{Shell weight of females}}{\text{Shell weight of males}} \times 100$$

Cocoons with individual values around 100 or more than 100 were selected for breeding during fixation of lines.

Table 11.1: Breeding of Sex Limited Silkworm Genotypes - Generation Effects (SL 7)

Generation	*Fecundity*	*Pupation female (%)*	*Pupation male (%)*	*Av. Cocoon wt.(g)*	*Single shell wt. (g)*	*Shell ratio (%)*
F-1	708	83.00	89.00	1.76	0.361	20.51
F-2	537	94.42	96.75	1.80	0.403	22.39
F-3	516	89.31	97.33	1.47	0.314	21.36
F-4	500	90.00	97.75	1.59	0.353	22.20
F-5	486	91.00	93.00	1.45	0.308	21.24
F-6	503	89.00	93.00	1.50	0.333	22.20
F-7	480	82.00	94.00	1.58	0.367	23.23
F-8	488	92.00	98.00	1.69	0.375	22.19
F-9	430	80.00	90.00	1.59	0.363	21.80

Table 11.2: Breeding of Sex Limited Silkworm Genotypes – Generation Effects (SL 8)

Generation	*Fecundity*	*Pupation female (%)*	*Pupation male (%)*	*Av. Cocoon wt.(g)*	*Single shell wt. (g)*	*Shell ratio (%)*
F-1	564	79.00	86.75	1.67	0.351	21.02
F-2	539	73.75	77.25	1.78	0.411	23.09
F-3	524	80.54	85.41	1.56	0.336	21.54
F-4	490	82.28	89.50	1.62	0.369	22.79
F-5	530	80.95	85.20	1.55	0.367	23.68
F-6	510	80.00	89.17	1.47	0.332	22.59
F-7	503	85.00	94.44	1.69	0.356	21.07
F-8	490	80.00	95.00	1.48	0.317	21.42
F-9	470	84.44	94.40	1.44	0.320	22.22

Table 11.3: Breeding of Sex Limited Silkworm Genotypes –Generation Effects (SL 9)

Generation	*Fecundity*	*Pupation female (%)*	*Pupation male (%)*	*Av. Cocoon wt.(g)*	*Single shell wt. (g)*	*Shell ratio (%)*
F-1	564	79.00	86.75	1.67	0.351	21.02
F-2	550	90.00	92.00	1.96	0.437	22.30
F-3	534	84.35	90.41	1.42	0.311	21.90
F-4	500	82.00	92.73	1.66	0.368	22.17
F-5	490	90.00	94.44	1.62	0.368	22.72
F-6	499	88.00	97.00	1.38	0.318	23.04
F-7	550	90.00	95.00	1.55	0.330	21.29
F-8	511	92.88	97.80	1.64	0.373	22.74
F-9	498	89.00	94.00	1.44	0.320	22.22

Table 11.4: Breeding of Sex Limited Silkworm Genotypes –Generation Effects (SL 10)

Generation	*Fecundity*	*Pupation female (%)*	*Pupation male (%)*	*Av. Cocoon wt.(g)*	*Single shell wt. (g)*	*Shell ratio (%)*
F-1	564	79.00	86.75	1.74	0.344	19.77
F-2	567	85.00	89.00	2.07	0.444	21.45
F-3	517	80.50	85.50	1.48	0.306	20.68
F-4	503	87.07	92.70	1.79	0.396	22.12
F-5	550	89.00	95.05	1.76	0.348	19.77
F-6	558	88.00	93.00	1.52	0.303	19.93
F-7	500	92.00	96.00	1.77	0.351	19.83
F-8	521	92.00	98.00	1.61	0.295	18.32
F-9	503	79.00	95.00	1.57	0.297	18.92

Table 11.5: Genetic Vigor Estimation in Sex Limited Races at F9 Generation against their mid Parent Value

Genotype	*Fecundity*	*Pupation Female (%)*	*Pupation Male (%)*	*Single cocoon wt. (g)*	*Single shell wt. (g)*	*Cocoon shell ratio (%)*
SL 7	(-) 15.69	(-) 1.23	(-) 2.17	(+) 12.67	(+) 21.36	(+) 7.72
SL 8	(-) 10.31	(+) 6.21	(+) 4.89	(-) 0.69	(+) 1.30	(+) 4.96
SL 9	(-) 4.96	(+) 12.66	(+) 4.44	(-) 0.69	(+) 4.23	(+) 4.96
SL 10	(-) 2.14	00	(+) 5.56	(+) 3.97	(-) 6.60	(-) 10.16

Table 11.6: Pooled Data of Female Specific Character of Breeding Lines (SL7, SL 8, SL9 and SL10) at F9.

Character	*Female*	*Male*
Pupation rate (%)	80.64	88.81
Single cocoon weight (g)	2.03	1.87
Single shell weight (g)	0.373	0.407
Cocoon shell ratio (%)	18.37	21.76
Cocoon filament length (m)	788	969
Denier	2.91	2.65

OBSERVATIONS

Donor parents of *Bombyx mori* L. viz., CC1 SL and NB4D2 SL had sexual dimorphism in cocoon colour having female cocoon with yellow colour and male with white. These races showed poor survival and also poor cocoon shell ratio during rearing. The productive silkworm parent races viz., O3 and D2 registered higher cocoon shell ratio between 21.78 to 24.16 percent, but poor survival during verification rearing. CC1 SL and O3 spun oval cocoons while NB4D2 SL and D2 spun dumbbell shaped cocoons. Females of sex-limited donor silkworm races were crossed with males of productive silkworm races to obtain F1 hybrids (Table 11.4).

Four silkworm hybrid combinations namely CC1SL x O3, CC1SL x D2, NB4D2 SL x O3 and NB4D2 SL x D2 were prepared and reared in mass batches. The resulting F1 cocoons exhibited 100% sexual dimorphism in cocoon colour. Female cocoons were golden yellow while male cocoons were white. 100 F1 cocoons from each hybrids population were randomly selected and subjected for inbreeding in F2 generation. Female cocoons exhibited 100% sexual dimorphism in cocoon colour. Female cocoon colour segregated into yellow and flesh (pinkish yellow) colour in F2 generation while males were white in colour. Flesh coloured female cocoons from CC1 SL x O3 hybrid population were separated and processed for inbreeding simultaneously along with yellow cocoon population in F3 generation. The resulting cocoons from F3 generation were selected and assessed individually for cocoon characters. Lines were drawn on the basis of colour of female cocoons, pupation rate and cocoon characters specially cocoons shell ratio. Individual single cocoon assessment was carried out at every generation from F3 to F7 generation. Selection of individual cocoons for continuation of next generation was resorted to on the basis of cocoon shell index as suggested by Yamamoto (1989). The cocoons with cocoon shell index values around 100 or more were selected for breeding. Egg layings with high fecundity were selected in every generation for next generation rearing.

Fecundity was high in F1 generation in all the four breeding populations which gradually decreased as generation progressed and was lowest between F7 to F9 generation. Fecundity ranged from 708 to 430 in SL7, 564 to 470 in SL8, 564 to 490 in SL9 and 567 to 500 in SL 10 genotypes. The pupation rate was comparatively low at F1 and F2 generations which gradually increased by every generation and was highest at F8 and F9 generations. Males revealed higher pupation rate than females.

Pupation rate in females varied from 80% to 94.42% in SL7, 73.75% to 85.00% in SL8, 79% to 92.88% in SL9 and 79 to 92% in SL10 genotypes. Pupation rate in males was as high as around 98% in preceding generations from F6 to F9. Cocoon weight in sex-limited lines was highest at F1 and F2 generations and gradually decreased in the preceding generations. Similar trend of gradual decrease in cocoon shell weight was recorded in all the sex limited lines except in case of SL7 where, shell weight decreased to the lowest level (0.308g) in F-5 generation and increased gradually in preceding generations. The cocoon shell ratio in sex limited genotypes ranged between 20.51 to 23.23% in SL7, 21.02 to 23.68% in SL8, 21.02 to 23.04 in SL9 and 18.32 to 22.12% in SL10 lines. Cocoon shell ratio in all the sex-limited lines was around 21% in F1 which slowly increased with few exceptions and was highest between F4 and F7 generation except in SL10 line, where shell ratio increased till F4 generation and decreased there after in preceding generation to the level of 18 to 19%.

GENETIC VIGOUR ESTIMATION IN SEX LIMITED GENOTYPES

Genetic vigour in sex limited lines was tested at F9 generation against the mid parental values of their respective parents of F1 generation (Table 11.5). All the sex-limited genotypes showed negative heterosis for fecundity character. Genotypes SL8, SL9, and SL10 recorded positive heterosis for pupation character in male and female larvae while SL7 genotypes recorded negative heterosis for pupation both in male and female larvae. Single cocoon weight showed positive heterosis in SL7 and SL10 genotypes. Single cocoon shell weight and cocoon shell ratio showed positive heterosis in SL7, SL8 and SL9 generations while SL10 recorded negative heterosis in both the characters.

Female larvae exhibited comparatively low pupation rate, low shell weight, poor cocoon shell ratio and higher single cocoon weight than males (Table 11.6). Male cocoons recorded higher cocoon filament length and thin denier as compared to females.

DISCUSSION

Tanaka (1916) was first to demonstrate that male of *Bombyx mori* L has ZZ homogametic sex chromosomes while female has ZW heterogametic sex chromosomes.

Z chromosome bears many genes responsible for morphological traits but no gene of morphological traits has been found on the W-chromosome.

Tazima (1941) using a special translocation from the 2nd chromosome to one end of the W-chromosome, which became marked by two noticeable genes +P and pSa, both for larval character provided strong evidence for monopolized role of W chromosome in sex determination of females. Female individuals were invariably found carrying + P and pSa genes of larval marking, indicating absence of crossing over between Z and W. The individuals carrying translocated W chromosome with dominant gene for larval marking were females with marked larvae.

Kimura *et al.* (1971) translocated the yellow blood gene (y) of the second chromosome on to the W- chromosome. The trasnslocated y genes co-existed with the cocoon colour genes (c) on the 12th chromosome, resulting production of yellow cocoons by females.

Present study envisaged low fecundity in sex-limited genotypes as compared to their parent genotypes. While other characters such as pupation rate, single cocoon weight, cocoon shell weight and cocoon shell ratio did not differ significantly in the sex-limited genotypes as compared to the parent races involved in breeding of these sex-limited genotypes. Rather, cocoon shell ratio improved in SL7, SL8 and SL9 genotypes from their mid parent values.

Our observations are in accordance with these observations and support the views of Yamamoto (1989) who opined that additional fragment of the chromosome attached to W sex-chromosome affects reproductive potential of females.

In our study, female cocoons recorded higher weight than males while male cocoons had higher cocoon shell weight and higher cocoon shell ratio. These observations substantiate the observations made by Tzenov (1993) who found that the female larvae utilize food effectively for body matter while males utilize food for silk production in sex-limited silkworm genotypes.

In present study, female moths of donor sex-limited genotypes of *Bombyx mori* L. were crossed with male moths of productive normal genotypes to prepare F1. The resultant F1 off springs became sex-limited after the transfer of 'W' chromosome with yellow gene (y) in females producing yellow coloured female cocoons and white cocoons in males. This sexual dimorphism became the genetic phenomenon as 'W' chromosome with yellow gene determined femaleness in every generation (Kimura *et al.*, 1971).

Kimura *et al.* (1971) while translocating yellow blood gene on to the 'W' chromosome translocated a large fragment containing I-lem gene locus (29.5) of the Gr gene locus (6.9) on second chromosome. This resulted lower cocoon weight and shell weight in females, poor fecundity and higher unfertilized eggs in sex-limited genotypes. Yamamoto (1989) suggested that females with higher cocoon shell weight than the males should be selected for breeding of sex-limited genotypes using cocoon shell index (shell wt. of females/shell wt. of male x 100), where shell index of 100 or more was used for selection in individuals for breeding.

In the present experiment, breeding was carried out as per the procedure suggested by Yamamoto (1989). However, the resulting genotypes at F9 generation recorded higher cocoon shell weight for male. But other economic character, except a slight decrease in fecundity remained on par with their productive parents. Rather, there was improvement in pupation rate in evolved sex-limited genotypes than their parents.

In the present observations on the breeding of sex limited silkworm genotypes on cocoon colour envisaged that viable sex limited genotypes can be evolved selecting higher fecundity, moderate cocoon weight (1.6 -1.7g) and higher cocoon shell ratio between 20–22% in every generation. During the study pupation rate especially in females was taken as first criteria for selecting batches or lines for further breeding. The heterosis cocoon weight and shell ratio in the genotypes were found negatively correlated with the heterosis pupation rate in females. Our observations also support the cocoon selection procedure on cocoon shell index suggested by Yamamoto (1989). This selection procedure can be effectively used in the breeding and also in maintenance of sex limited breeds at P4 and P3 levels.

REFERENCES

Kimura, K., Harada, C. & Akai, H. (1971). Studies on W chromosomes translocation of yellow blood gene in silkworm. *Jpn. J. Breed.*, 2(4): 199-203.

Mundukar, R., Murthy, M., Guptha K.N.N., Rao, K.S. & Raghuraman, R. (2002), Sex-limited races of silkworm races, *Bombyx mori*, as useful grainage tools. *Indian Silk,* 41(2): 13-15.

Tanaka, Y. (1916). Genetic studies in the silkworm. *J. Coll. Assoc. Sapporo,* 7: 129-155.

Tazima, Y. (1941). A simple method of sex discrimination by means of larval markings in *Bombyx mori* L. *J. Seric. Sci. Jpn.,* 12: 184-188.

Tzenov, P. (1993). Study on the food utilization in genetic sex-limited breeds for eggs and larval traits of the silkworm. *B. mori* at moderate, reduced and excess feeding amounts. *Sericologia,* 33(2): 247-256.

Yamamoto, T. (1989). Breeding of sex limited yellow cocoon races of silkworms by chromosome manipulation. *Farming Japan,* 23(5): 42-48.

12

MEASURES TO IMPROVE THE SILKWORM COCOON PRODUCTIVITY IN AUTUMN AND INTRODUCTION OF SUMMER CROP IN NORTH-WESTERN STATES

Abad A. Siddiqui, M.A. Khan, and Mir Nisar Ahmad

ABSTRACT

For competing in international market and to sustain the sericulture industry, India should produce more of bivoltine cocoons by utilizing the opportunities available in northwestern states especially in Jammu & Kashmir, Uttar Pradesh, Uttarakhand and Himanchal Pradesh in the form of salubrious climate, better soil conditions, irrigation facilities etc. Since southern states are producing mainly cross-breed (Multi. × Biv.) cocoons being the tropical areas, therefore, for quality bivoltine cocoon production with silk filament of above 2Agrade, the climatic conditions of above states ,which are well suited to bivotline rearing, should be utilized properly with meticulous management and planning. In U.P., Uttrakhand and Himanchal Pradesh only two bivoltine crops are taken in a year at farmers' level, one in spring and another in autumn season and Jammu & Kashmir sericulture still thrives only on one spring crop. Further, the success rate of silkworm crop during monsoon/autumn season is very low and the cocoon productivity is also poor. It is one of the biggest challange to different agencies who are involved with the development of sericulture that the opportunities bestowed by the nature to this region should be utilized properly. There are several reasons of low productivity in autumn seasons and

Central Sericultural Research & Training Institute, Pampore-181101, J&K., India.

E-mails: siddiquibad@yahoo.w.co.in; csrtipper@vsnl.com; mirnisarahmad@gmail.com non introduction of summer crop in these states, such as non-availability of hardy bivoltine silkworm breeds, non-adoption of advanced package of practices for maintenance of mulberry wealth, non adoption of new rearing technologies etc. In the present paper, problems faced by north western states in development of sericulture industry and the measures to be taken to improve the productivity in autumn season and also for the introduction of one additional crop in summer season have been discussed in detail.

Key words: Sericulture, management, summer crop, improvement

SERICULTURE SCENARIO IN NORTHWESTERN STATES

In Jammu and Kashmir, only one bivoltine crop is in vogue in spite of salubrious climate most suitable for silkworm rearing. In other state such as U.P., Uttrakhand, Himanchal Pradesh and sub-topical parts of Jammu only two bivoltine silkworm crops are being taken in spring and autumn seasons and the productivity level in autumn is only around 25-30 kg/100 dfls. In autumn, there is frequent reports of crop losses due to different silkworm diseases. In Jammu and Kashmir presently about 21000 families are engaged in practicing sericulture contributing to the production of 762 MT of cocoons. The present raw silk production is 102 MT., In spite of the fact that climatic conditions are suitable for bivoltine rearing and also being the traditional state, only one silkworm crop is taken at farmers level. In recent years, autumn crop was also introduced that too after great motivation, but only 10% of total rearers involved in autumn rearing and the productivity is around 10 kg/ /100 dfls.

Uttar Pradesh, major raw silk consuming state of the country, is consuming about 2500 – 3000 MT raw silk. The state is known for manufacturing of world famous Banarsi silk sarees. Mulberry sericulture is practiced both on nature grown mulberry trees and plantation raised. 827 ha of land has been brought under mulberry plantation. While the production of cocoons during 2006-07 has been 85 tons with production of raw silk to the tune of 25 tons. Under non-mulberry sector, Uttar Pradesh has got 950 ha tropical tasar plantation with a raw silk production of 8 tons. Eri culture is introduced in Kanpur Dehat and 100 farmers are involved in this practice. About 3085 families are engaged in sericulture activities in the state.

Uttarakhand state has potential of 140 tons of cocoon production per year. The mulberry raw silk production in 2006-07 was14 metric tons. Under non mulberry sector major production is from Oak sector while as Eri & Muga are in infant stage. 668 hectares of land is available as mulberry wealth & and 3000 families are engaged in sericulture activities. Scattered plantation of Som, Soalu & Castor are available but systematic plantation of these plants are required to be taken up.

Himanchal Pradesh has congenial climate well suited for production of quality bivoltine silk cocoons. The sericulture activities are being practiced in eight districts viz. Hamirpur, Bilaspur, Mandi, Kangra, Shimla, Solan, Sirmur and Una. Development of sericulture in hill state is most advantageous because it improves ecology of fragile Himalyan mountains. At present, about 8000 families are practicing sericulture in the state. A total of1685 hectares of land is presently under mulberry plantation in the state. During the year 2006-07 raw silk preproduction in the state was 17 MT .

CONSTRAINTS IN DEVELOPMENT OF SERICULTURE

Remuneration generated from sericulture (out of two sericulture crop in a year), is not very lucrative and rather uneconomical. Due to that, farmers do not pay much time to sericulture activities and they concentrate more to other agriculture crops . This is posing a major problem to different states and central govt. agencies to convince the farmers that the sericulture is more profitable than other agriculture crops. Govt. agencies are also involved to motivate the farmers to improve and increase their mulberry plantation and to contribute more time to sericulture crop. Low return is coming in big way to peruse the farmers to adopt new sericulture technologies for improving the productivity per unit area of the land. Farmers of the area have developed a mind set that that adoption of new technologies will increase their expenditure on sericulture and will further make the sericulture more uneconomical. They think that the expenditure of even a single penny on sericulture and occupation of their prime land under mulberry cultivation will be a waste and not going to increase their return from sericulture. Even they are reluctant to grow mulberry as tree on the periphery of their field. Since farmers are not paying any attention to the mulberry plantation in way of application of farm yard manure, chemical fertilizers, timely pruning etc. due to that, whatever leaves available on scattered mulberry trees are of low nutritive value and this is also making productivity level low in north-western states. The average cocoon yield per oz. of seed is around 30-35 kg against the national average of 45 kg. per 100 dfls. For making the sericulture economically viable, maximization of bivoltine crop in a year is very essential. Simultaneously, i) Strengthening of extension net work ,ii) Introduction of new mulberry cultivation technologies iii) Introduction of improved rearing technologies with maintenance of strict hygiene and iv) Adoption of new silkworm disease control measures are essential to improve the productivity level in these states.

WHY LOW PRODUCTIVITY IN AUTUMN AND REASONS OF NON-INTRODUCTION OF BIVOLTINE SUMMER CROP ?

In north-western states temperature starts shooting up from May onwards and remains towards higher side up to August (30-40°C)) and again from November temperature starts decreasing and remain very low during winter months up to February. This situation is forcing the farmers to restrict to only to two silkworm crop.

There are mainly following three reasons of low productivity in autumn and also of non- introduction of summer crop in north-western states:

1) Low quality of mulberry leaves during the period.
2) Non adoption of proper rearing technologies suitable for summer and autumn seasons and
3) Non-availability of sturdy bivoltine hybrids to tolerate high temperature and high humid conditions during summer and autumn months.

SUGGESTIVE MEASURES TO ADDRESS THE PROBLEMS

i) Improvement of quality of mulberry leaves

Most of the farmers in the region are having local mulberry varieties , which need to be replaced in phase manner with improved mulberry varieties . A lot of plantation was done by govt. agencies under different schemes. However, because of low return from sericulture due to aforesaid reasons farmers have not attended the plantation and consequently a large hectare of mulberry has been vanished or remained in very poor condition.

Another reason of low quality of mulberry in autumn season is that the farmers are not attending proper pruning schedule. In north-western states, spring rearing is completed around 5th of April. After that sprouting started slowly and a tree again attains a productive stage with full flush of leaf around 15th of May (after 45 days of spring crop). What is happening in the field that the farmers are neither taking any crop on these leaves in May nor they prune the garden as recommended in last weak of June or first weak of July. Due to that, leaves which have sprouted after spring crop become 4 months old and very coarse with lot of leaf mulberry diseases especially leaf spot . These leaves are not suitable for rearing and cause low productivity as well as feeding of coarse leaves with dust and diseases also increase the percentage of crop loss.

If a crop is introduced in May then farmers will utilise the leaves which are available in May -June and after that farmers can attend the pruning during last weak of June or first weak of July. Consequently, fresh 45 days old good quality leaves will be available in autumn, suitable for silkworm rearing. Feeding of fresh diseased free leaves will improve the productivity and will also reduce the percentage of crop loss in autumn. To increase the productivity in autumn very sincere efforts are to be made to motivate the farmers to adopt pruning package. Leaves around 15th May after spring crop on same mulberry plantation if not utilized in time (May-June) will become unfit up to autumn rearing . Therefore, summer crop should be introduced to utilize the leaf in time and also to make sericulture more remunerative.

ii) High temperature and high humid conditions

Another reason for low yield in autumn and non-introduction of summer crop is the prevailing high temperature and high humid conditions. To cope up the problems of high temperature and high humidity following measures are suggested:

a) Modifications in rearing practices and strict maintenance of hygiene

It is a general observation that rearing practices are very poor at farmers level and improper management during rearing decreases the productivity, therefore, to improve the productivity in autumn and summer crop following steps are suggested which should be ensured.

1. For the success of cocoons crop, it is utmost important to make the rearing environment free from pathogens. If the rearing environment is free from pathogen, no out break of diseases during rearing will be observed. Proper and strict modification drive should be ensured by govt. agencies. Chinese system of mass disinfection drive by utilizing co-operative agencies should be adopted. This strategy is not only advocated for unfavorable seasons but also to be adopted for spring crop as well.

2. Strangers should not be allowed inside the rearing room.
3. Hands and feet to be washed before cleaning of feeding operation.
4. Never handle larvae with bare hands and always use chopsticks.
5. Net cleaning should be regular. Nets once used should be disinfected and dried before every use.
6. Do not transfer the larvae on the floor of the rearing house.
7. Use vinyl sheet on the floor of the rearing bed.
8. Do not preserve leaves inside the rearing room.
9. Spilling rearing bed waste or keeping leaves meant for feeding on the floor to be avoided.
10. As the nutrition plays a major role in silkworm rearing, utmost care has to be bestowed to keep up the leaf quality throughout the period of rearing. Succulent leaves with higher moisture content are ideally suited for young age. Raising robust larvae will minimize loss of silkworms during latc age.
11. Regular use ofbed disinfectant should be a part of the integrated disease management schedule.
12. Bed disinfectant cannot provide desired results if application is delayed till the appearance of diseases in the rearing bed. Hence, it should be used as per schedule through out the year.
13. Proper ventilation of rearing room during rearing and spinning should be ensured to avoid accumulation of toxic gases in the rearing houses.

b) Development of hardy breeds

Bivoltine breeds/hybrids developed so far are suitable for spring season and whenever they are reared in unfavourable seasons of summer they succumb to different diseases. Therefore, sturdy breeds tolerant to high temperature and humidity and suitable for rearing under prevailing poor rearing practices at farmers level should have to be developed to introduce one or two additional

Results of Field trials of ATR16 X ATR 29 along with control

LOCATION	DATE OF BRUSHING	DFLS REARED	YIELD/100 DFLS (Kg)
Himanchal Pradesh	1) 10-05-03	100	35.00
	2) 10-05-05	350	59.85
	3) 10-09-05	100	45.00
	Control		40.0
	4) 10-05-06	300	39.1
	Control	300	35.0
U.P.	25-08-05	50	80.0
	Control	50	40.0
Uttarakhand	1)15-05-03	100	40.0
	Control	100	5.0
	2) 20-05-04	400	43.6
	Control	500	42.0
	3) 24-05-05	550	17.0
	Control	600	8.0
	24-05-06	500	30.5
	Control	500	12.2

crops in a year. What is the main reason that silkworm hybrids developed by different institute till recently, are not suitable to poor rearing practices and high temperature and high humidity because during the development of new breeds, breeders, as one of the breeding strategy, maintained optimum environmental conditions during filial generations, though with this strategy a number of productive bivoltine hybrids were developed but they are not suitable for rearing in unfavourable seasons. According to Chinese breeders , hardy breeds should be developed by exposing the breeding lines during fixation to high temperature and high humidity and should be fed with low quality mulberry leaves, so as resultant breeds suitable to these conditions. RSRS, Sahaspur has recently developed hardy bivoltine hybrid ATR16 x ATR29 for unfavourable season. New hybrid was tested along with control during summer and autumn in Himanchal Pradesh in summer in Dehradun area and in autumn in U.P. 2003,2004 2005 and 2006.In Himanchal Pradesh in all testing, brushing was taken on 10th of May while during testing in Dehradun area brushing were taken on 10 and 20th of May.

Following inferences have been drawn based on the observation made during trial of ATR hybrid.

1. The average cocoon yield /100 dfls in ATR hybrid was 32.2 kg and in control the yield was only 17 kg.
2. It is observed that average cocoon yield/100 dfls was about 20% more in trials taken on 10th of May than average cocoon yield obtained from the trials which were taken after 25th of May. It may be due to the reason that in June number of hotter days (room temperature more than 33°C) are more than the May. Though in May and June there was not much difference in maximum temperature but still May is comparatively more suitable because number of days with maximum temperature in May are less than June. Therefore, it is recommended that the summer rearing may be introduced form 15th of May in Uttarakhand, U.P. ,Himanchal Pradesh . In U.P since the

spring crop is over by last weak of April and the fresh flush of leaves will be available on same patch of plantation by 15th of April (after 45 days of spring rearing), therefore, summer rearing may be taken from 15th of April.

3. It is a also observed that state department in both U.P. and Uttarakhand take additional staggered crop in April in small scale on the left over leaves after first spring crop utilizing the same hybrid recommended particularly for spring season. In place of spring specific hybrid, hardy bivoltine hybrids may be utilized for staggered crop.

However, it will be more appropriate that all leaves available in spring should be utilized fully in spring itself, for that extra facilities, rearing space , rearing trays etc. may be available at farmers level. Summer crop may be introduced as additional crop not as staggered crop on fresh flush available after 40-45 days of Spring crop to increase the return. In U.P. 2-3 Multi x Bivoltine crops are in practice in summer and autumn. In autumn also state is taking Multi x Bivoltine rearing, multi x Bi rearings may be continued in summer and monsoon. However, in autumn Bi x Bi rearing with utilizing ATYR hybrid may be started.

Though, in Uttarakhand autumn crop is almost established and only improvement in productivity has to be achieved with minimizing percentage of crop loss . Contrary to this, in Jammu and Kashmir second crop is still not fully established. Now efforts have been made by CSR&TI, Pampore along with DOS Kashmir for establishing autumn crop in J&K, which enjoys the comparatively better climate in autumn also for bivoltine cocoon crop.

EFFECTS OF MULBERRY (*MORUS* SP.) TRANSPIRATION SUPPRESSANTS ON ECONOMIC TRAITS OF SILKWORM (*BOMBYX MORI* L.)

*B. K. Singhal, Anil Dhar, B. B. Bindroo and M. A. Khan**

ABSTRACT

The present study deals with favourable influences on silkworm economic traits by using transpiration suppressants, namely, abscisic acid, 8-HQ, cycocel, kaolin and salicylic acid on mulberry during spring and autumn seasons in Indian sub tropics. During spring season, the weight of 10 mature silkworm larvae was found increased by 1.89% by using 8-HQ, cycocel and kaolin as compared with control. However, single cocoon and shell weights were found increased by 10.65 and 19.32%, respectively by the use of cycocel followed by 7.10 and 15.93% increase by using 8-HQ as compared with control. However, salicylic acid has increased single cocoon and shell weights by 5.92 and 13.90%, respectively and its use was found economically viable than using 8-HQ, cycocel and kaolin. But, during autumn season, instead of cycocel and 8-HQ, salicylic acid was found to improve the silkworm economic traits. Weight of 10 mature silkworm larvae was found increased by 7.32% by using salicylic acid. Single cocoon and shell weights were also increased by 3.22 and 4.02% by the use of this chemical as compared with control. Therefore, the present findings, though, revealed the use of cycocel during spring and salicylic acid during autumn season, but, overall salicylic acid is suggested for both the seasons as most economically viable transpiration suppressant in mulberry genotype Chinese White for influencing silkworm economic traits favourably in Jammu region of Indian sub tropics.

Regional Sericultural Research Station, Miransahib, Jammu-181101.
Central Sericultural Research & Training Institute, Pampore-192121, J&K., India.
E-mail: csrtipper@vsnl.com

Key words: Mulberry, Transpiration suppressants, Abscisic acid, 8-HQ, Cycocel, Kaolin, Salicylic acid, Silkworm economic traits.

INTRODUCTION

The silkworm cocoon crop in Jammu region of Indian subtropics has so far been commercialized only for the spring season (March-April), though, second crop during autumn season (September-October) is also coming up. This situation in this region is prevailing due to extreme climatic conditions. Therefore, being a profession only for 2 to 3 months in a year, mulberry could not compete with other agricultural crops in the region and farmers are reluctant in cultivation of mulberry plantations in their own cultivated fields. As such, farmers grow mulberry only as border trees in their agricultural field or in backyards of their houses or as roadside trees exclusively under rainfed conditions, that too, without attending training of trees. As a result, the mulberry leaf is always in short supply with poor quality under such adverse conditions (Singhal *et al.,* 2003, 2005). During rearing period, farmers bring leaf from far-flung areas which looses lot of leaf moisture and deteriorates its quality for silkworm feeding. The increase in mulberry leaf productivity by the use of plant hormones has been reported by Nanja Reddy and Prasad (1999). The favourable effects of cycocel have been reported by Rajat Mohan *et al.* (2005) on mulberry grown under different adverse situations *i.e.* higher alkalinity and rainfed conditions to improve water status of mulberry. They found increase in absolute silk recovery/10,000 larvae by 57% to 68% when silkworms were fed with cycocel treated mulberry leaves. The plant growth regulators are reported for their favourable effects on productivity in various crops including mulberry (Biswas and Sengupta, 1993; Malik, 1998). In sericulture, it is the mulberry leaf quality which contributes maximum (38.2%) than any other factor for successful silkworm cocoon crop (Miyashita, 1986). Therefore, keeping in view the poor leaf quality of untrained mulberry trees grown under rainfed conditions of Jammu region, the present study was aimed to exploit transpiration suppressants for improving mulberry leaf quality for influencing silkworm economic traits favourably and results are highlighted in the present communication.

Materials and Methods

The mulberry transpiration suppressants namely, abscisic acid ($C_{15}H_{20}O_4$), 8-HQ (8- Hydroxy quinoline sulphate), cycocel (2-Chloroethyl trimethyl ammonium chloride), kaolin and salicylic acid were foliar sprayed during spring and autumn seasons on mulberry plants of the genotype Chinese White grown in the experimental field of Regional Sericultural Research Station, Jammu. The 5 years old bush type of plantation was used for the study raised under 3 x 3' spacings. For spray of these mulberry transpiration suppressants their stock solutions were prepared by dissolving 8-HQ and salicylic acid in 5 ml. of ethanol and then diluting them with distilled water. Cycocel and kaolin were dissolved in distilled water. Abscisic acid was dissolved in 5 ml of 1 N NaOH and then diluted with distilled water. Two foliar sprays of abscisic acid, 8-HQ, cycocel, kaolin and salicylic acid at their concentrations of 0.50 ppm, 2.00 ppm, 2.00 ppm, 5.00% and 50.00 ppm, respectively were undertaken during spring and autumn seasons. First spray @ 500 liters of solution per hectare was taken up at the onset of foliage after pruning. Second spray of 600 liters solution per hectare was taken up after one week of first spray. Tween 20 (0.01% concentration) was used as surfactant. For estimation of leaf quality, chlorophyll "a", "b" and carotenoides were estimated by the DMSO method of Hiscox and Israelstam (1979). The methods of AOAC (1980) were followed for the estimation of crude protein and carbohydrates. The ascorbic acid was estimated by the method of Ranganna (1977). For their influence on silkworm economic traits, rearing was conducted with silkworm hybrid $SH_6 \times NB_4D_2$ by feeding mulberry leaf sprayed with transpiration suppressants after III moult. Influence on silkworm economic traits were observed for their possible exploitation under prevailing harsh rainfed conditions of Jammu region. The experiment was conducted for three years in spring and autumn seasons and average data have been presented.

Results and Discussion

The influences of different mulberry transpiration suppressants on leaf quality and silkworm economic traits during spring and autumn seasons are presented in Tables 13.1 to 13.6. The mulberry leaf quality varies between both the seasons and gets influenced by the use of transpiration suppressants through stomatal pores. As a result, the leaf turgidity maintained by checking excessive water loss which regulates physiological and biochemical processes favourably. As is indicated Chlorophyll "a", Chlorophyll "b", carotenoids, protein, carbohydrates and ascorbic acid contents have been effected favourably in both the seasons by the use of ascorbic acid, 8-HQ, cycocel, kaolin and salicylic acid. Chlorophyll "a", Chlorophyll "b"and carotenoid contents were higher in the mulberry plants sprayed with cycocel followed by 8-HQ. Protein content was increased by 16.79% by cycocel spray followed by 16.56% and 16.33% by the spray of 8-HQ and salicylic acid, respectively during spring season. Regarding influence of such improved leaf quality on silkworm economic traits, the single cocoon weight was found increased by 10.65%, 7.10% and 5.92% by the use of cycocel, 8-HQ and salicylic acid respectively, during spring season. Similarly, increase in single shell weight was also recorded by 19.32%, 15.93% and 13.90% by the use of cycocel, 8-HQ and salicylic acid respectively. The transpiration suppressants cycocel, 8-HQ and salicylic acid have increased protein content sharply and the same may be the reason for improvement in silkworm economic traits. Fukuda *et al.* (1960) reported that about 70% of the silk protein produced by the silkworm is directly derived from the proteins of the mulberry leaf. The protein content of the mulberry leaves has a direct correlation between the production efficiency of cocoon shell in silkworm and

Table 13.1: Influence of mulberry suppressants on silkworm economic trait during autumn season

Mulberry leaf quality parameter	Mulberry transpiration suppressants					
	Control	Abscisic acid (0.50 ppm)	8-HQ (2.0 ppm)	Cycocel (2.0 ppm)	Kaolin (5 %)	Salicylic acid (50.00 ppm)
Chlorophyll "a" (mg/g)	1.95	1.99	2.30	2.38	1.99	2.28
Chlorophyll "b" (mg/g)	0.94	1.05	1.21	1.23	1.01	1.20
Carotenoids (mg/g)	1.23	1.32	1.50	1.57	1.32	1.57
Protein (%)	17.45	19.01	20.34	20.38	18.71	20.30
Carbohydrates (%)	9.42	9.41	9.62	9.63	8.71	9.60
Ascorbic acid (%)	1.83	1.89	1.92	1.92	1.81	1.90

Table 13.2: **Influence of mulberry transpiration suppressants on silkworm economic traits during spring season.**

Mulberry transpiration suppressant	Concentration (ppm)	Larval duration (Day:Hr.)	Weight of 10 mature larvae (g)	Yield/10,000 larvae brushed		Single cocoon wt. (g)	Single shell wt. (g)	SR %
				By no.	By wt.			
Control		25 : 06	53.00	9650	19.25	1.69	0.295	17.45
Abscisic acid acid	0.50	25 : 00	50.00	9600	19.90	1.76	0.331	18.81
8-HQ	2.00	25 : 00	54.00	9500	19.45	1.81	0.342	18.89
Cycocel	2.00	25 : 00	54.00	9500	19.35	1.87	0.352	18.82
Kaolin	5 %	25 : 05	54.00	9600	18.50	1.67	0.285	17.06
Salicylic acid	50.00	25 : 05	53.00	9500	18.90	1.79	0.336	18.77

Table 13.3: **Mulberry leaf protein content *vs* single cocoon and single shell weight as influenced by cycocel, 8-HQ and salicylic acid ranked at top three during spring season.**

% increase in parameter as compared with control	Cycocel	8-HQ	Salicylic acid
Mulberry leaf protein	16.79	16.56	16.33
Single cocoon weight	10.65	7.10	5.92
Single shell weight	19.32	15.93	13.90

Table 13.4: **Influence of transpiration suppressants on mulberry leaf quality during autumn season.**

Mulberry leaf quality parameter	Mulberry transpiration suppressants					
	Control	Abscisic acid (0.50 ppm)	8-HQ (2.0 ppm)	Cycocel (2.0 ppm)	Kaolin (5 %)	Salicylic acid (50.00 ppm)
Chlorophyll "a" (mg/g)	1.91	2.16	2.14	1.94	2.16	2.18
Chlorophyll "b" (mg/g)	0.92	1.17	1.15	1.13	1.16	1.18
Carotenoids (mg/g)	1.23	1.52	1.29	1.29	1.53	1.54
Protein (%)	16.21	19.06	18.83	18.99	19.05	19.36
Carbohydrates (%)	8.31	9.04	8.41	8.43	9.04	9.11
Ascorbic acid (%)	1.72	1.82	1.77	1.77	1.82	1.86

the mulberry leaf protein content as reported by Machii and Katagiri,(1991). As mulberry leaf nutrients have a role to play in silkworm nutrition, the optimum use of nutrients becomes inevitable (Singhal and Mala, 1998; Singhal *et al.*, 2001). But, during autumn season, leaf quality was found performed better by the use of salicylic acid followed by kaolin. By the application of salicylic acid, the most crucial nutrient of sericulture i.e. protein content was found increased by 19.43% followed by 17.52% increase by kaolin as compared to control in autumn season. By the application of salicylic acid, the single cocoon and shell weights were found increased by 3.22 and 4.02%, respectively as compared with control. Rajat Mohan *et al.* (2005) also found favourable effects of cycocel on mulberry grown under different adverse situations *i.e.* higher alkalinity and rainfed conditions to improve water status of plants. While observing the influence of cycocel on silkworm economic traits, they have further reported that the absolute silk recovery/10,000 larvae can be increased by 57 to 68% when silkworms are fed with cycocel treated mulberry leaves. Singhal *et al.* (2006) reported that salicylic acid can be of use for increasing leaf yield and quality in the promising mulberry variety S-146 commonly used for silkworm rearing by the farmers under rainfed conditions of Jammu province with improvement in silkworm economic parameters at highly profitable cost benefit ratio's of 1 : 19.9 and 1 : 19.7 during spring and autumn seasons. However, Singhvi *et al.* (2001) found that the response of salicylic acid is very much specific to mulberry variety as its application increases the leaf yield by 4.71% in the variety Tr-10 to 15.17% in S-36. Therefore, to tackle the scenario of sericulture under such extreme climatic conditions under rainfed conditions of Jammu region affecting mulberry leaf quality is ultimately leading to poor silkworm economic traits. The finding reveals that out of five transpiration suppressants, salicylic acid during autumn and, cycocel during spring season are suited for improvement in silkworm economic traits by using Chinese White as the mulberry variety. However, salicylic acid is suggested

Table 13.5: **Influence of mulberry transpiration suppressants on silkworm economic traits during autumn season.**

Mulberry transpiration suppressant	Concentration (ppm)	Larval duration (Day:Hr.)	Weight of 10 matured larvae (g)	Yield/10,000 larvae brushed		Single cocoon wt. (g)	Single shell wt. (g)	SR %
				By no.	By wt.			
Control		26 : 04	41.00	9500	14.45	1.55	0.298	19.22
Abscisic acid acid	0.50	26 : 03	45.00	9600	13.65	1.56	0.298	19.10
8-HQ	2.00	26 : 02	40.00	9350	13.35	1.52	0.294	19.34
Cycocel	2.00	26 : 04	39.00	9350	13.75	1.55	0.290	18.72
Kaolin	5 %	25 : 04	41.00	9500	14.50	1.58	0.299	18.92
Salicylic acid	50.00	25 : 04	44.00	9500	14.70	1.60	0.310	19.37

Table 13.6: **Mulberry leaf protein content *vs* single cocoon and single shell weight as influenced by salicylic acid, abscisic acid and kaolin ranked at top three during autumn season.**

% increase in parameter as compared with control	Cycocel	8-HQ	Salicylic acid
Mulberry leaf protein	19.43	17.58	17.52
Single cocoon weight	3.22	0.64	1.93
Single shell weight	4.02	Nil	0.33

for both the seasons, when economic viability is concerned to boost sericulture particularly in Jammu region.

ACKNOWLEDGEMENTS

The authors are highly thankful to Mr. B. D. Mandi, Technical Assistant, for his help in many ways.

REFERENCES

Association of Official Analytical Chemists (1980). *Methods of analysis.* 13th ed., Washington, D.C.

Biswas, S. & Sengupta, K. (1993). Effect of hormones on the mulberry – A review. *Sericologia,* 33: 461-472.

Fukuda, T. (1960). The correlation between the mulberry leaves taken by the silkworms, the silk protein in the silk gland and the silk filament. *Bull. Sericult. Exp. Stn.,* 15: 595-610.

Hiscox, J.D. & Israelstam, G..F. (1979). A method for the extraction of chlorophyll from tissue without maceration. *Can. J. Bot.,* 57: 1332-1334.

Machii, H. & Katagiri, K. (1991). Varietal differences in nutritive values of mulberry leaves for rearing silkworms. *JARQ,* 25: 202-208.

Miyashita, V. (1986). A report on mulberry cultivation and training methods suitable to bivoltine rearing in Karnataka. Central Silk Board, Bangalore, India, 27: 99-104.

Malik, C. P. (1998). *Advances in plant hormones research: Indian scenario.* Agro Botanica, Bikaner, India, p. 303.

Reddy, N.Y.A. & Prasad, T.G. (1999). Potential application of growth regulators in mulberry. In: *Advances in mulberry sericulture.* (eds. Devaiah, M.C., Narayanswamy, K.C. and Maribashetty, V.G.) CVG Publications, Bangalore, India, pp. 123-144.

Mohan, R., Juyal, A.C., Ramakant, Singh, P.K. and Singh, B.D. (2005). Response of foliar spray of CCC on growth and quality of mulberry. In: *Advances in tropical sericulture.* (eds. Dandin, S.B., Mishra, R.K., Gupta, V.P. and Reddy, Y. S.) NASSI, Bangalore, pp. 164-165.

Ranganna, S. (1977). *Manual of analysis for fruits and vegetable products.* Tata McGraw Hill Pub. Ltd., New Delhi, India.

Singhal, B.K., Dhar, A., Bindroo, B.B., Bakshi, R.L. & Khan, M.A. (2003). Sericulture practices and future strategies under present scenario of Indian subtropics. *Int. J. Indust. Entomol.,* 7: 107-115.

Singhal, B.K., Dhar, A., Tripathi, P.M., Qadri, S.M.H., Saxena, N.N., Bindroo, B.B. & Khan, M.A. (2005). Leaf nutritional quality in different plantation types of mulberry (Morus sp.) in Indian sub-tropics. In: *Advances in tropical sericulture.* (eds. Dandin, S.B., Mishra, R.K., Gupta, V.P. and Reddy, Y. S.) NASSI, Bangalore, pp. 153-156.

Singhal, B.K., Dhar, A., Khan, M.A., Sengupta, D., Bindroo, B.B. & Dhar, S.L.. (2006). Technological approach of using salicylic acid as an anti-transpirant for increasing crop productivity of mulberry (*Morus* spp.) under rainfed condition of Indian sub-tropics. In: *Appropriate Technologies for Mulberry Sericulture in Eastern and North Eastern India.* Proc. of the Workshop held during 17-18 January, 2006 at CSR & TI, Berhampore, India, pp.53-58.

Singhal, B.K., Dhar, A., Qadri, S.M.H. & Ahsan, M.M. (2001). Mulberry nutrition for development of sericulture in Jammu and Kashmir. *Asian Textile J.,* 10: 35-42.

Singhal, B.K. & Mala, V.R. (1998). An insight into silkworm's food. *The Indian Textile J.,* 108: 86-88.

Singhvi, N.R., Kodandaramaiah, J., Rekha, M., Sarkar, A. & Datta, R.K. (2001). Effect of salicylic acid on leaf yield and pigment content in mulberry (*Morus* spp.). *Indian J. Seric,.* 40: 100-102.

IDENTIFICATION OF PROMISING BIVOLTINE HYBRIDS OF SILKWORM *BOMBYX MORI* L.

M. Ramesh Babu, H. Lakshmi, J. Prasad, J. Seetharamulu and Chandrashekharaiah

ABSTRACT

Sixteen newly developed bivoltine hybrids of silkworm *Bombyx mori* L. at Andhra Pradesh State Sericulture Research and Development Institute (APSSRDI) Hindupur, are evaluated for eleven economic traits contributing to silk yield by adopting two selection methods *viz*., Evaluation Index and Subordinate Function for their overall and relative economic merit. The hybrid genotype APS_{11} x APS_{6} and its reciprocal combination that ranked first scoring highest values in Evaluation Index (64.97 and 59.93) and Subordinate Function (9.4089 and 8.3298) methods are adjudicated as most promising and recommended for commercial utilization. Since, the performance of reciprocal combination is on par with the direct cross *i.e.* APS_{11} x APS_{6}, it is highly advantageous for the graineurs in terms of increase in egg production and considerable reduction in cost of egg production.

Key words: Bivoltine silkworm, Hybrids. Evaluation Index. Subordinate Function, Direct and Reciprocal cross

INTRODUCTION

Sustained and increased quality silk production depends on the utilization of the optimal potential of the hybrids of silkworm *Bombyx mori* L. The silkworm with many economic traits is best exemplified for exploitation of hybrid vigour. Generally, the hybrids are characterized with greater vigour, faster growth, better

Andhra Pradesh State Sericulture Research and Development Institute (APSSRDI) Kirikera – 515 211. Hindupur (Ananthapur District. Andhra Pradesh).

productivity, higher tolerance to diseases and unfavourable climatic conditions (Ashoka and Govindan, 1990). Although, there is quantitative increase in the overall silk production in India over years (Rao *et al*., 2001), there remains ample need for potential silkworm hybrids. Considering the obvious limitations with regard to productivity and quality of raw silk from cross breed type that constitute core raw silk production more emphasis was laid on to popularize bivoltine sericulture under tropical/sub tropical conditions of India. Various successful efforts are made especially during the last decade in the development of productive bivoltine breeds (Datta *et al*., 2000a, 2000b, 2000c & 2001 and Chandra-shekharaiah and Ramesh 2003). Since, the genetic improvement of multiple traits being the objective of evolving the productive bivoltine hybrids suitable for Indian, tropical climate, many breeders (Begum *et al*., 2000, Babu *et al*., 2001, 2002 and Rao *et al*., 2001) followed the Evaluation Index method of Mano *et al*., 1993) to adjudicate the superior hybrids. Further, Babu *et al*. (2001 & 2002) and Rao *et al*. (2001) also adapted an alternate approach *i.e*., Subordinate Function method (Gower, 1971) for evaluation of comprehensive merit and subsequent identification of promising bivoltine hybrids. In the present investigation, an attempt has been made to evaluate the overall merit of the newly developed bivoltine hybrids at APSSRDI, Hindupur and subsequent identification of most promising bivoltine hybrid(s) for commercial exploitation by following both Evaluation Index and Subordinate Function methods.

MATERIALS AND METHODS

Thirty two bivoltine hybrid genotypes (sixteen each of oval x peanut and peanut x oval type) involving eight parents namely, APS_{11}, APS_5, APS_{13} and APS_7 (oval) and APS_6, APS_8, APS_{18} and APS_2 (peanut) developed at APSSRDI, Hindupur, constituted the study materials. These new hybrid genotypes were reared along with the two control bivoltine hybrids *viz*., APS_9 x APS_8 (Kalpatharuvu) and APS_5 x APS_4 (Hemavathy). All the combinations were brushed as composite in three replications each. Each composite laying comprising about 1000 – 1200 eggs were represented by 10 – 12 different broods. 300 larvae after III moult were retained per replication breedwise and the rearings were conducted as per the standard method (Krishnaswamy, 1976 and Bhargava *et al.*, 1993). The data were collected for eleven economic traits *viz.*, fecundity, cocoon yield per 10000 larvae by weight, pupation rate, cocoon weight, shell weight, shell ratio, absolute silk content, filament length, raw silk recovery, reelability and neatness and analyzed by following the Evaluation Index and Subordinate Function methods as detailed below.

Evaluation Index Method (Mano *et al.*, 1993)

The Evaluation Index values character wise were calculated by using the formula

$$\text{Evaluation Index (E.I.)} = \frac{A - B}{C} \times 10 + 50$$

Where,

A = Value obtained for a trait for a hybrid

B = Mean of a trait of all the hybrids for the trait

C = Standard deviation of all the hybrids for a trait

10 = Standard unit

50 = Fixed value

The E.I. values obtained character wise and combination wise were added and average cumulative E.I. value over the number of traits studied was derived. These values were arranged in descending order. The hybrids with average E.I. value above 50 were considered to possess economic merit. The hybrid with highest average E.I. was adjudicated to be the most promising. Considering the advantage of utilizing the reciprocal combinations for egg production, average E.I. values for both direct and reciprocal crosses together were also calculated and adjudicated the promising combination that possess the highest average E.I.

Subordinate Function Method (Gower, 1971)

The Subordinate Function values character wise for each of the combination were added and cumulative Subordinate Function value over the number of traits was calculated using the following formula:

$$Xu = (Xi - X\ min)/\ (X\ max - X\ min)$$

Where,

Xu	=	Subordinate Function
Xi	=	Measurement of character of a tested genotype
X max	=	The maximum value of the character from all the tested genotypes
X min	=	The minimum value of the character among all the tested genotypes

The values are arranged in descending order and the combination with highest cumulative Subordinate Function value was assigned first rank and subsequent ranks were arranged in descending order.

RESULTS AND DISCUSSION

The greater and sudden climatic fluctuations coupled with poor quality mulberry leaf and low management by the farmers under tropical conditions warrant more flexible genotypes, so as to achieve sustainable increased silk production. The superiority of the silkworm hybrid genotype is judged by more than 21 economic traits (Thiagarajan *et al.*, 1993). Eleven silk yield and quality contributory traits were considered and analysed in the present study to identify the most promising hybrid (s) from 32 newly developed bivoltine hybrid genotypes.

The mean performance of the 32 hybrid genotypes is presented in Table 14.1. The data present that the performance of the newly

Table 14.1: Mean performance of new hybrids

Sl. No.	Hybrid Combination	Fecundity (No.)	Cocoon.yld by weight (kg)	Pupation Rate (%)	Cocoon weight (g)	Cocoon Shell weight (g)	Shell Ratio (%)	Filament Length (m)	Raw silk recovery (%)	Reelability (%)	Neatness (%)	Absolute Silk Content (kg)
DIRECT CROSSES												
1	APS13 x APS2	446	14.6	91.8	1.619	31.6	19.5	816	14.3	75.7	82.3	2.848
2	APS13 x APS18	441	15.0	93.9	1.574	32.4	20.6	841	14.7	76.2	83.7	3.094
3	APS13 x APS8	536	15.2	93.4	1.580	31.7	20.1	815	15.1	74.8	85.3	3.049
4	APS13 x APS6	442	15.8	94.7	1.615	35.3	21.8	850	15.4	76.0	83.7	3.442
5	APS5 x APS2	521	18.6	90.4	2.008	42.5	21.2	838	15.3	76.4	84.7	3.934
6	APS5 x APS18	505	17.3	92.0	1.861	40.0	21.5	798	15.2	75.8	89.0	3.729
7	APS5 x APS8	526	18.4	93.8	1.943	45.6	23.5	785	17.4	80.8	89.0	4.320
8	APS5 x APS6	546	18.4	92.8	1.938	41.7	21.5	922	14.9	75.5	86.3	3.962
9	APS11 x APS2	542	18.9	91.6	1.899	40.6	21.4	726	15.3	75.3	85.7	4.040
10	APS11 x APS18	525	18.1	92.4	1.906	41.2	21.6	777	15.9	75.6	89.7	3.923
11	APS11 x APS8	501	17.6	90.7	1.894	40.4	21.3	768	15.6	76.0	90.7	3.743
12	APS11 x APS6	516	18.7	92.3	2.053	49.1	23.9	909	18.4	80.9	87.3	4.465
13	APS7 x APS2	509	17.1	91.6	1.854	38.4	20.7	778	15.4	76.3	78.0	3.546
14	APS7 x APS18	505	17.8	91.3	1.896	39.8	21.0	831	15.0	75.8	82.3	3.733
15	APS7 x APS8	474	16.9	91.9	1.800	39.0	21.7	791	16.8	75.6	84.7	3.662
16	APS7 x APS6	496	17.2	92.1	1.843	38.8	21.1	786	15.1	78.3	83.0	3.629

Contd....

RECIPROCAL CROSSES											
17 APS2 x APS13	481	16.3	90.1	1.696	32.8	19.3	679	13.2	74.2	86.0	3.144
18 APS2 x APS5	477	16.0	91.6	1.722	36.0	20.9	620	13.7	68.4	85.3	3.349
19 APS2 X APS11	461	16.4	90.6	1.797	36.5	20.3	704	15.8	72.6	84.7	3.337
20 APS2 x APS7	488	16.3	91.4	1.837	38.0	20.7	735	14.9	76.6	85.7	3.368
21 APS18 x APS13	456	17.5	89.6	1.913	40.8	21.3	801	16.3	80.0	87.3	3.739
22 APS18 x APS5	475	16.9	89.4	1.860	42.4	22.8	885	16.3	76.9	86.7	3.848
23 APS18 x APS11	536	17.4	89.5	1.937	42.8	22.1	836	15.8	76.3	89.0	3.855
24 APS18 x APS7	520	18.4	91.3	1.883	38.7	20.5	729	15.3	75.1	83.7	3.770
25 APS8 x APS13	567	17.2	90.2	1.977	44.4	22.4	773	16.0	78.6	88.0	3.861
26 APS8 x APS5	506	18.4	91.1	1.903	42.1	22.1	785	17.4	80.8	89.0	4.066
27 APS8 x APS11	553	18.9	90.7	1.988	41.6	20.9	897	16.8	76.1	89.0	3.966
28 APS8 x APS7	497	17.3	91.3	1.924	39.1	20.3	723	14.1	75.2	84.7	3.510
29 APS6 x APS13	523	18.0	90.4	1.962	42.2	21.5	728	16.1	72.9	86.3	3.877
30 APS6 x APS5	467	18.4	90.3	1.987	42.5	21.1	930	17.2	76.5	91.0	3.882
31 APS6 x APS11	577	19.5	93.8	2.182	46.2	21.2	920	16.0	74.5	84.3	4.134
32 APS6 x APS7	530	16.1	90.3	1.775	40.0	22.5	870	16.1	79.4	85.3	3.623
APS5 x APS4	506	13.8	90.9	1.583	32.3	20.4	785	13.5	76.9	85.9	2.815
APS9 x APS8	489	14.5	92.5	1.506	31.6	21.0	803	14.1	80.0	87.6	3.045
Average	504	17.1	91.5	1.845	39.4	21.3	801	15.5	76.4	86.0	3.7
S.D	34.8	1.42	1.34	0.15	4.41	0.99	72.61	1.16	2.54	2.72	0.40
CV%	6.90	8.27	1.46	8.38	11.21	4.64	9.06	7.48	3.32	3.16	10.94

developed hybrids exhibited superiority in different individual characters and no single hybrid combination excelled in all the traits analyzed. Further, it is also evident that the coefficient of variation is within the acceptable limit except for the higher variation observed for the traits cocoon shell weight (11.2%) followed by absolute silk content (10.94%). The hybrids showed higher survival above 91.5% which is very important under tropical conditions. These observations corroborate with Vidyunmala *et al.* (1998) and Babu, *et al.* (2002) who concluded that superiority in one or a couple of characters may not reflect the overall superior merit of the hybrid.

In the Evaluation Index method, out of the 16 combinations, 8 combinations *viz.*, APS_{11} x APS_6, APS_5 x APS_8, APS_5 x APS_6, APS_{11} x APS_{18}, APS_5 x APS_2, APS_{11} x APS_2, APS_5 x APS_{18} and APS_{11} x APS_8 were found to possess economic merit with cumulative index value above 50 (Table 14.2). APS_{11} x APS_6 with a highest value of 64.97 ranked first. Among the reciprocal combinations, the genotype APS6 x APS11 recorded highest value of 59.93. The lowest Evaluation Index value of 38.40 was recorded for APS_{13} x APS_2. Based on the average cumulative index, the hybrid combinations were ranked in descending order and APS_{11} x APS_6 occupied the first place. On the other hand, the two control hybrids recorded E.I. values of 44.06 (APS_9 x APS_8) and 40.11 (APS_5 x APS_4) respectively.

In the Subordinate function method, the cumulative values obtained for the 16 direct and their reciprocal combinations are presented in Table 14.3. The scrutiny of the data present that the cumulative values among the direct crosses varied from a maximum of 9.4089 (APS_{11} x APS_6) to a low of 2.6095 (APS_{13} x APS_2) while among the reciprocal combinations the maximum and minimum values recorded were 8.3298 (APS_6 x APS_{11}) and 2.6702 (APS_2 x APS_{13}) respectively. The control hybrids, APS_9 x APS_8 and APS_5 x APS_4 recorded Subordinate Function values of 3.9832 and 3.0135 respectively.

Table 14.2: **Evaluation index values for new hybrids**

Sl. No.	Hybrid Combination	Fecundity	Cocoon.yld by weight	Pupation Rate	Cocoon weight	Cocoon Shell weight	Shell Ratio	Filament Length	Raw silk recovery	Reelability	Neatness	Absolute Silk Content	Avg. E.I.
1	APS13 x APS2	33.37	31.83	52.12	35.38	32.49	32.34	52.11	39.11	47.42	36.42	29.79	**38.40**
2	APS13 x APS18	31.74	35.12	67.59	32.47	34.23	42.85	55.55	42.53	49.52	41.32	35.94	**42.62**
3	APS13 x APS8	59.24	36.29	64.34	32.90	32.64	37.51	51.92	46.45	44.01	47.46	34.81	**44.33**
4	APS13 x APS6	32.22	40.29	74.07	35.14	40.73	55.52	56.70	48.86	48.74	41.32	44.65	**47.11**
5	APS5 x APS2	54.83	60.04	41.89	60.59	57.20	48.92	55.09	47.57	50.31	45.01	56.95	**52.58**
6	APS5 x APS18	50.23	51.11	53.62	51.06	51.46	52.09	49.58	47.08	47.82	60.96	51.82	**51.53**
7	APS5 x APS8	56.27	58.87	67.09	56.37	64.15	72.23	47.79	66.00	67.52	60.96	66.61	**62.17**
8	APS5 x APS6	62.02	59.10	59.60	56.02	55.23	52.16	66.66	44.70	46.77	51.14	57.66	**55.55**
9	APS11 x APS2	60.96	62.39	50.62	53.54	52.82	50.84	39.71	48.32	45.85	48.69	59.60	**52.12**
10	APS11 x APS18	56.08	56.98	56.36	53.99	54.25	53.38	46.64	53.42	46.90	63.41	56.67	**54.37**
11	APS11 x APS8	48.99	52.99	43.64	53.22	52.29	50.15	45.40	50.69	48.61	67.09	52.17	**51.38**
12	APS11 x APS6	53.39	60.98	55.86	63.48	72.08	76.58	64.87	74.60	67.92	54.70	70.23	**64.97**
13	APS7 x APS2	51.48	49.70	50.37	50.63	47.90	44.43	46.83	48.54	49.92	20.47	47.25	**46.14**
14	APS7 x APS18	50.33	54.40	48.13	53.35	51.08	47.11	54.17	45.76	47.69	36.42	51.92	**49.12**
15	APS7 x APS8	41.22	48.28	52.62	47.09	49.19	53.86	48.57	61.07	47.03	45.01	50.15	**49.46**
16	APS7 x APS6	47.64	50.64	54.61	49.90	48.74	47.54	47.88	46.62	57.54	38.87	49.31	**49.03**

Contd....

17 APS2 x APS13	43.43	43.82	39.40	40.36	35.06	30.13	33.15	30.14	41.51	49.91	37.18	**38.55**
18 APS2 x APS5	42.09	41.94	50.37	42.05	42.46	46.32	25.07	34.27	18.78	47.46	42.31	**39.37**
19 APS2 X APS11	37.58	44.99	42.89	46.92	43.52	40.11	36.68	51.93	35.20	45.01	42.03	**42.44**
20 APS2 x APS7	45.44	43.82	49.38	49.49	46.92	44.01	40.95	44.22	50.84	48.69	42.78	**46.05**
21 APS18 x APS13	36.15	52.75	35.91	54.45	53.27	50.32	49.99	56.83	64.50	54.82	52.07	**51.01**
22 APS18 x APS5	41.70	48.05	34.41	51.02	56.97	65.36	61.61	56.26	52.02	52.37	54.81	**52.23**
23 APS18 x APS11	59.05	52.05	34.91	56.00	57.88	58.24	54.81	51.96	49.66	60.96	54.96	**53.68**
24 APS18 x APS7	54.45	58.63	48.13	52.51	48.43	42.31	40.12	47.66	45.06	41.32	52.85	**48.32**
25 APS8 x APS13	68.15	50.40	40.40	58.54	61.35	61.63	46.18	54.25	58.72	57.27	55.12	**55.64**
26 APS8 x APS5	50.52	58.87	46.88	53.78	56.22	58.16	47.79	66.00	67.52	60.96	60.25	**57.00**
27 APS8 x APS11	63.93	62.63	43.64	59.30	55.16	46.41	63.22	60.84	48.87	60.96	57.74	**56.61**
28 APS8 x APS7	48.03	50.87	48.13	55.12	49.42	40.21	39.25	37.34	45.58	45.01	46.35	**45.94**
29 APS6 x APS13	55.50	56.28	41.89	57.62	56.44	51.97	39.89	54.82	36.26	51.14	55.53	**50.67**
30 APS6 x APS5	39.40	58.87	40.65	59.21	57.12	47.95	67.81	64.00	50.71	68.32	55.65	**55.43**
31 APS6 x APS11	70.93	66.63	67.09	71.83	65.51	49.05	66.43	53.53	42.56	43.78	61.95	**59.93**
32 APS6 x APS7	57.51	42.64	40.65	45.52	51.38	62.32	59.50	54.54	62.14	47.46	49.18	**52.08**
APS5 x APS4 ©	50.52	26.42	45.39	33.07	34.00	40.95	47.79	32.46	52.15	49.55	28.96	**40.11**
APS9 x APS8 ©	45.63	31.36	57.36	28.09	32.42	47.02	50.27	37.62	64.37	55.80	34.72	**44.06**

The peer analysis of the data of both direct and reciprocal combinations of the newly developed hybrid genotypes put together present that the combination APS_{11} x APS_6 that scored the highest E.I. value of 62.45 and ranked first. Similarly, the same combination ranked first with highest value of 8.8694. Further, the relative ranking of the remaining combinations in both the methods is similar suggesting the applicability to identify the promising hybrids as has already observed by Babu, *et al*. (2002) and Rao *et al.* (2001).

Further, the superiority expressed in terms of percent improvement of the new hybrid, APS_{11} x APS_6 over the control hybrids *viz.*, APS_5 x APS_4 and APS_9 x APS_8 is presented in Table 6. The data present that the new hybrid showed highest percent improvement of 58.61 and 46.63 and 52.01 and 55.38 for the traits absolute silk content and cocoon shell weight over APS_5 x APS_4 and APS_9 x APS_8 respectively.

Unlike in the one way of egg preparation in multivoltine x bivoltine combinations, wherein the female component of bivoltine and male component of multivoltine breeds are wasted, in bivoltine x bivoltine hybrid egg production the hybrid genotypes of both direct and reciprocal combinations could be used as the variation in the over all performance is very less. This enhances not only egg production efficiency but also contribute to grainage economics immensely. Considering this, the average cumulative values and relative merit of both direct and reciprocal combinations of all the 16 newly developed hybrid combinations were derived at (Tables 4 (Evaluation Index) and 5 Subordinate Function) establishing the superiority of APS_{11} x APS_6 and its reciprocal combination.

A few silkworm breeders (Singh and Subba Rao 1993 and Udupa and Gowda, 1988) used selection indices for identification of promising hybrids based on only few characters. Since, the comprehensive merit of the hybrids depends on various viable, quantitative and qualitative traits, selection need to be based on multiple trait analysis that warrants reliable methods exercising due

Table 14.3: Subordinate Function values for new hybrids

Sl. No.	Hybrid Combination	Fecundity (No.)	Cocoon.yld by weight	Pupation Rate	Cocoon weight	Cocoon Shell weight	Shell Ratio	Filament Length	Raw silk recovery	Reel-ability	Neat-ness	Abs.Silk Content	Cumulative index value
1	APS13 x APS2	0.0416	0.1345	0.4465	0.1667	0.0019	0.0475	0.6327	0.2018	0.5829	0.3333	0.0201	2.6095
2	APS13 x APS18	0.0000	0.2164	0.8365	0.1001	0.0457	0.2737	0.7132	0.2785	0.6257	0.4359	0.1692	3.6948
3	APS13 x APS8	0.7017	0.2456	0.7547	0.1100	0.0057	0.1589	0.6284	0.3669	0.5134	0.5641	0.1416	4.1909
4	APS13 x APS6	0.0122	0.3450	1.0000	0.1612	0.2095	0.5466	0.7401	0.4210	0.6096	0.4359	0.3801	4.8613
5	APS5 x APS2	0.5892	0.8363	0.1887	0.7431	0.6248	0.4043	0.7025	0.3920	0.6417	0.5128	0.6781	6.3134
6	APS5 x APS18	0.4719	0.6140	0.4843	0.5251	0.4800	0.4726	0.5736	0.3810	0.5909	0.8462	0.5539	5.9936
7	APS5 x APS8	0.6259	0.8070	0.8239	0.6464	0.8000	0.9063	0.5317	0.8066	0.9920	0.8462	0.9121	8.6981
8	APS5 x APS6	0.7726	0.8129	0.6352	0.6386	0.5752	0.4742	0.9731	0.3275	0.5695	0.6410	0.6954	7.1153
9	APS11 x APS2	0.7457	0.8947	0.4088	0.5819	0.5143	0.4458	0.3426	0.4088	0.5508	0.5897	0.7425	6.2256
10	APS11 x APS18	0.6210	0.7602	0.5535	0.5922	0.5505	0.5005	0.5048	0.5235	0.5722	0.8974	0.6713	6.7472
11	APS11 x APS8	0.4401	0.6608	0.2327	0.5745	0.5010	0.4310	0.4758	0.4623	0.6070	0.9744	0.5624	5.9218
12	APS11 x APS6	0.5526	0.8596	0.5409	0.8092	1.0000	1.0000	0.9313	1.0000	1.0000	0.7154	1.0000	9.4089
13	APS7 x APS2	0.5037	0.5789	0.4025	0.5153	0.3905	0.3078	0.5091	0.4139	0.6337	0.0000	0.4432	4.6986
14	APS7 x APS18	0.4743	0.6959	0.3459	0.5774	0.4705	0.3655	0.6810	0.3514	0.5882	0.3333	0.5562	5.4397
15	APS7 x APS8	0.2421	0.5439	0.4591	0.4344	0.4229	0.5109	0.5499	0.6957	0.5749	0.5128	0.5133	5.4598
16	APS7 x APS6	0.4059	0.6023	0.5094	0.4985	0.4114	0.3747	0.5338	0.3707	0.7888	0.3846	0.4931	5.3734
17	APS2 x APS13	0.2983	0.4327	0.1258	0.2806	0.0667	0.0000	0.1890	0.0000	0.4626	0.6154	0.1992	2.6702

18 APS2 x APS5	0.2641	0.3860	0.4025	0.3190	0.2533	0.3484	0.0000	0.0928	0.0000	0.5641	0.3235	2.9537
19 APS2 X APS11	0.1491	0.4620	0.2138	0.4305	0.2800	0.2148	0.2718	0.4900	0.3342	0.5128	0.3166	3.6756
20 APS2 x APS7	0.3496	0.4327	0.3774	0.4892	0.3657	0.2988	0.3716	0.3166	0.6524	0.5897	0.3349	4.5786
21 APS18 x APS13	0.1125	0.6550	0.0377	0.6026	0.5257	0.4345	0.5832	0.6003	0.9305	0.7179	0.5600	5.7599
22 APS18 x APS5	0.2543	0.5380	0.0000	0.5242	0.6190	0.7584	0.8550	0.5874	0.6765	0.6667	0.6263	6.1057
23 APS18 x APS11	0.6968	0.6374	0.0126	0.6381	0.6419	0.6052	0.6960	0.4907	0.6283	0.8462	0.6300	6.5232
24 APS18 x APS7	0.5795	0.8012	0.3459	0.5582	0.4038	0.2622	0.3523	0.3939	0.5348	0.4359	0.5787	5.2463
25 APS8 x APS13	0.9291	0.5965	0.1509	0.6963	0.7295	0.6780	0.4941	0.5422	0.8128	0.7692	0.6337	7.0324
26 APS8 x APS5	0.4792	0.8070	0.3145	0.5873	0.6000	0.6034	0.5317	0.8066	0.9920	0.8462	0.7582	7.3259
27 APS8 x APS11	0.8215	0.9006	0.2327	0.7135	0.5733	0.3505	0.8926	0.6905	0.6123	0.8462	0.6973	7.3310
28 APS8 x APS7	0.4156	0.6082	0.3459	0.6179	0.4286	0.2169	0.3319	0.1618	0.5455	0.5128	0.4214	4.6065
29 APS6 x APS13	0.6064	0.7427	0.1887	0.6750	0.6057	0.4701	0.3469	0.5551	0.3556	0.6410	0.6438	5.8311
30 APS6 x APS5	0.1956	0.8070	0.1572	0.7115	0.6229	0.3837	1.0000	0.7614	0.6497	1.0000	0.6467	6.9357
31 APS6 x APS11	1.0000	1.0000	0.8239	1.0000	0.8343	0.4072	0.9678	0.5261	0.4840	0.4872	0.7994	8.3298
32 APS6 x APS7	0.6577	0.4035	0.1572	0.3984	0.4781	0.6929	0.8056	0.5487	0.8824	0.5641	0.4899	6.0784
APS5 x APS4 ©	0.4792	0.0000	0.2767	0.1139	0.0400	0.2329	0.5317	0.0522	0.6791	0.6077	0.0000	3.0135
APS9 x APS8 ©	0.3545	0.1228	0.5786	0.0000	0.0000	0.3636	0.5897	0.1683	0.9278	0.7385	0.1394	3.9832

Table 14.4: **Average Evaluation Index values for direct and reciprocal hybrids**

Sl. No.	Hybrid Combination	Direct Cross	Reciprocal Cross	Average E.I.
1	APS11 x APS6	64.97	59.93	62.45
2	APS5 x APS8	62.17	57.00	59.58
3	APS5 x APS6	55.55	55.43	55.49
4	APS11 x APS18	54.37	53.68	54.03
5	APS5 x APS2	52.58	39.37	45.98
6	APS11 x APS2	52.12	42.44	47.28
7	APS5 x APS18	51.53	52.23	51.88
8	APS11 x APS8	51.38	56.61	54.00
9	APS7 x APS8	49.46	45.94	47.70
10	APS7 x APS18	49.12	48.32	48.72
11	APS7 x APS6	49.03	52.08	50.55
12	APS13 x APS6	47.11	50.67	48.89
13	APS7 x APS2	46.14	46.05	46.09
14	APS13 x APS8	44.33	55.64	49.98
15	APS13 x APS18	42.62	51.01	46.82
16	APS13 x APS2	38.40	38.55	38.47
	APS9 x APS8 ©	44.06		44.06
	APS5 x APS4 ©	40.11		40.11

Table 14.5: **Average Subordinate function values for direct and reciprocal hybrids**

Sl. No.	Hybrid Combination	Direct Cross	Reciprocal Cross	Average SF value
1	APS11 x APS6	9.4089	8.3298	8.8694
2	APS5 x APS8	8.6981	7.3259	8.0120
3	APS5 x APS6	7.1153	6.9357	7.0255
4	APS11 x APS18	6.7472	6.5232	6.6352
5	APS5 x APS2	6.3134	2.9537	4.6336
6	APS11 x APS2	6.2256	3.6756	4.9506
7	APS5 x APS18	5.9936	6.1057	6.0496
8	APS11 x APS8	5.9218	7.3310	6.6264
9	APS7 x APS8	5.4598	4.6065	5.0331
10	APS7 x APS18	5.4397	5.2463	5.3430
11	APS7 x APS6	5.3734	6.0784	5.7259
12	APS13 x APS6	4.8613	5.8311	5.3462
13	APS7 x APS2	4.6986	4.5786	4.6386
14	APS13 x APS8	4.1909	7.0324	5.6116
15	APS13 x APS18	3.6948	5.7599	4.7273
16	APS13 x APS2	2.6095	2.6702	2.6398
	APS9 x APS8 ©	3.9832		3.9832
	APS5 x APS4©	3.0135		3.0135

weightage to all the economic characters. For such selection methods various index methods are followed for both plants and live stock breeding programmes (Arunachalam and Bandyopadhyay, 1984, Begum *et al.*, 2000, Babu *et al.*, 2002, Rao *et al.*, 2001 and Vidyunmala *et al.*, 1998). In silkworm breeding the Evaluation Index method as detailed by Mano *et al.*, (1993) is being widely applied and the same is adapted in the present study also. The analysis presents that the hybrid genotype APS_{11} x APS_6 together with its reciprocal combination is assigned first. Further, these findings are ensured with the Subordinate Function index (Gower, 1971) method that is relatively applied by very few breeders (Babu *et al.*, 2002, Rao *et al.*, 2001 and Rao *et al.*, 2003). The combination APS_{11} x APS_6 and its reciprocal together recorded highest value of 8.8694 and ranked first. The subsequent ranks assigned to the hybrids are similar in both the methods. Further, the hybrid genotype APS_{11} x APS_6 showed the desired higher percent improvement over the control hybrids APS_5 x APS_4 and APS_9 x APS_8 establishing its superiority.

Therefore, the new hybrid APS_{11} x APS_6 that ranked first in both the methods along with its reciprocal combination is adjudicated as most promising hybrid and recommended for commercial utilization. Since, it's reciprocal combination is also on par with the APS_{11} x APS_6, it is highly advantageous for the graineur in terms of increased egg productivity and considerable reduction in the cost of egg production.

REFERENCES

Arunachalam, V. & Bandyopadyay, A. (1984). A method to make decision jointly on a number of independent characters. *Indian. J. Genet.*, 44: 419-424.

Ashoka, J. & Govindan, R. (1990). Heterosis for pupal and related traits in single and double cross hybrids of bivoltine silkworm, *Bombyx mori* L. *Entomon.*, 15: 203 – 206.

Bhargava, S.K., Thiagarajan, V., Babu, M.R. & Majumdar, M.K. (1993). Impact of silkworm hybrids on reeling parameters. *Textile J.*, 104: 77-89.

Chandrashekharaiah & Babu, M.R. (2003). Silkworm breeding in India during the last five decades and what next? A lead paper presented in Mulberry Silkworm Breeders Summit held at APSSRDI, Hindupur. 6-13.

Datta, R.K., Basavaraja H.K., Reddy, N.M., Kumar, S.N., Ahsan, M.M., Kumar, N.S. & Babu, M.R. (2000a). Evolution of new productive hybrids CSR_2 x CSR_4 and CSR_2 x CSR_5. *Sericologia*, 40:151 – 167.

Datta, R.K., Basavaraja, H.K., Reddy, N.M., Kumar, S.N., Ahsan, M.M., Kumar, N.S., & Babu, M.R. (2000b). Evolution of new productive bivoltine hybrid, CSR_3 x CSR_6. *Sericologia,* 40: 407-416.

Datta, R.K., Kumar, N.S., Basavaraja, H.K. & Reddy, N.M. (2000c). CSR_{18} x CSR_{19} – A robust bivoltine hybrid for rearing throughout the tropics. *Current Technology Seminar on Sericulture Technology.* An appraisal 7 – 7, June, CSRTI, Mysore.

Datta, R.K., Basavaraja, H.K., Reddy, N.M., Kumar, S.N., Kumar, N.S., Babu, M.R., Ahsan, M.M. & Jayaswal, K.P. (2001). Breeding of new productive bivoltine hybrid, CSR_{12} x CSR_6 of silkworm *Bombyx mori* L. *Int. J. Indust. Entomol.*, 3: 127 – 133.

Gower, J.C. (1971). A general coefficient of similarity and some of its properties. *Biometrics,* 27: 857-871.

Krishnaswamy, S. (1976). New Technology of Silkworm Rearing. Bulletin No. 2, Central Silk Board, Bangalore.

Mano, Y., Kumar, S.N., Basavaraja, H.K., Reddy, N.M. & Datta, R.K. (1993). A new method to select promising silkworm breeds/combinations. *Indian Silk*, 31: 53.

Begum, A.N., Basavaraja, H.K., Rao, P.S., Rekha, M. & Ahsan, M.M. (2000). Identification of bivoltine silkworm hybrids suitable for tropical climate. *Indian J. Seric.*, 30: 24-29.

Babu, M.R., Lakshmi, H.C. & Prasad, J. (2001). Silkworm (*Bombyx mori* L.) genetic stocks – an evaluatory analysis. *Bull. Indian Acad. Seric.*, 5: 9-17.

Babu, M.R., Lakshmi, H.C. & Prasad, J. (2002). Multiple Trait Evaluation of Bivoltine Hybrids of Silkworm (*Bombyx mori* L.). *Int. J. Indust. Entomol.*, 5: 37-43.

Rao, C.G.P., Basha, K.C.I., Seshagiri, S.V., Ramesh, C. & Nagaraju, H. (2003). Identifiication of superior polyvoltine hybrids (polyvoltine x bivoltine) of silkworm, *Bombyx mori* L. *Int. J. Indust. Entomol.*, 8: 43-49.

Singh, T. & Rao, G..S. (1993). Multiple Trait evaluation Index to select useful silkworm (*Bombyx mori* L.) hybrid genotypes. *Indian Entomol.*, 6:370-382.

Rao, P.S., Singh, R., Kalpana, G..V.., Nishitha, Naik, V., Basavaraja H.K., Swamy G.N.R. & Datta, R.K. (2001). Evaluation and identification of promising bivoltine hybrids of silkworm, *Bombyx mori* L. for tropics. *Int. J. Indust. Entomol.*, 3: 31-35.

Thiagarajan, V., Bhargava, S.K., Babu, M.R. & Nagaraju, B. (1993). Differences in seasonal performance of twenty six strains of silkworm, *Bombyx mori* L. (Bombycidae). *J. Lepidopter. Soc. America,* 47: 331-337.

Udupa, S. & Gowda, B.L.V. (1988). Heterotic expression in silk productivity of different crosses of silkworm (*Bombyx mori* L.). *Sericologia*, 28: 395-400.

Vidyunmala, S., Murthy, B.N. & Reddy, N.S. (1998). Evaluation of new mulberry silkworm (*Bombyx mori* L.) hybrids (multivoltine x bivoltine) through multiple trait evaluation index. *J. Entomol. Res.*, 22: 49-53.

STUDIES ON THE RESIDUAL EFFECT OF BOTANICAL PESTICIDES ON SILK WORM, *BOMBYX MORI* L. AND ITS ECONOMIC PARAMETERS

S.K. Mukhopadhyay, M.V. Santha Kumar, P. Mitra, S.K. Das and A.K. Bajpai

ABSTRACT

Effective concentrations of three botanical insecticides viz.. Neem oil, Pongamia oil and [Nicotine extract were independently and in combinations evaluated critically for their residual effect on silkworm.*Bombyx mori* L. to derive the safe/ waiting period. Botanical insecticide sprayed leaves were fed to silkworms immediately after spray, 5 days, 7 day. 14 days and 21 day of spraying from hatching onwards till initiation of spinning. Rearing data revealed that 1.5% Pongamia oil, 2% Nicotine extract, and 1% Neem oil after 14 days of spray have shown the survivability on par with all the economic characters with untreated control. Therefore, botanical insecticides can safely be used for mulberry crop protection 14 days after spray without having any impact on the economic traits of silkworm.

Key words: Mulberry. Pests, Botanical pesticides, Residual toxicity, Silkworm,Safe period..

INTRODUCTION

Mulberry *(Morus* spp.) grows abundantly in the eastern & northeastern region of India due to prevalence of hot and humid condition and adequate rainfall except during brief winter months. Among several factors that influence mulberry leaf yield insect pest

Entomology Laboratory. Central Sericultural Research & Training Institute (CSR&TI). Berhampore-742101. W.B.
E-mails: csrtiber@rediffmail.com: akbjp@rediffmail.com

and non-insect pest form an important component. Among the pests that attack mulberry three are major viz. Thrips, Mealy bug & Whitefly. They are responsible from 11 - 24% crop loss during March -November and affect the quality that ultimately influence the production of silk. Several authors have worked with botanicals on different pests in different crops (Flint & Parks, 1989, Gupta & Lai, 1998, Gupta & Rawat, 2004, Mann *et al.*, 2001, Nath *et.al.*, 2002, Patel *et al.,* 1990, Sundararaj *et al.,* 1995, Heyde *et al.*, 1984, Nimbalkar *et al.*, 1990, Pillai 1988, Asoka & Patil 2001, Ishman 1999). Neem formulations are widely used in whitefly and thrips management (Walter 1998, Sharma *et al.,* 1999, Ishan 1999, Sattar 2001, Mukhopadhyay 2005, 2006). Tobacco extracts were found effective against sucking pests (Tesfaye & Gautam, 2003) Pongamia oil can be isolated in water medium and acts as an insecticidial agent (Chakravorty *et al., 1976,* Chandrika Mohan *et al.*, 2001). Botanical insecticides treated leaves were tested with silkworms (Bandyopadhyay *et al.,* 2002 & 2005) and residual toxicity/waiting period was ascertained. Nine concentrations and combinations of three botanical insecticides viz. Neem oil. Pongatnia oil & Nicotine extract were found effective in management of major mulberry pests like Thrips, mealy bug and whitefly (Mukhopadhyay *et al.,* 2006, 2000a).

Several chemical insecticides were recommended to contain the pest population below ETL. But chemical controls often pose a problem due to their longer waiting period as time period between the spray and feeding to the silk worm is less, several botanical insecticides were tried on mulberry pests (Bandyopadhyay 2002, Mukhopadhyay 2006, 2006 a) and found successful in controlling pests.

In this study, investigations were made to study the optimum waiting/safe period of botanicals sprayed leaves that can be fed to silkworm without having detrimental effect on the economic characters.

MATERIALS & METHODS

Mulberry leaves were sprayed with different concentrations and combinations viz. 1.0% Neem oil, 1.5% Pongamia oil, 2.0%

Pongamia oil, 1.0% nicotine extract, 2.0% Nicotine extract, 1.0% Nicotine extract + Pongamia oil (1 : 1), 1.0% Neem oil+ Pongamia oil (1:1), 1.0%Nicotine ex. + Pongamia oil (1:1) and 1.0% Neem oil + Pongamia oil (10:1) along with recommended chemical (0.1% dimehoate and 0.1% DDVP) for Thrips, mealy bug and whitefly with an unsprayed control. The leaves from the sprayed plots were fed to the silkworms from the 7 Day, 14th day, and 21st day of spray from hatching to spinning. Mortality was recorded day wise. After spinning cocoons were subjected to assessment of cocoons and reeling analysis for their economic parameters.

RESULTS & DISCUSSION

Treated leaves from nine treatments were fed to silkworms. Silkworm lots fed with mulberry leaves 5 DAS (days after spray) mortality in silkworms was observed.. When leaves fed with mulberry leaves 7 DAS, no significant mortality was observed (9.67 — 4%). Maximum mortality (9.67%) was observed when 1.0% Nicotine ex. + Pongamia oil (1:1), 8.0% mortality was observed in 1.5% Pongamia oil treated leaves and minimum mortality was observed with 1.0% Neem oil and 2.0% Pongamia oil sprayed leaves were fed. Among the treatments 2% Pongamia oil has shown maximum S.R.% (15.09%) followed by 1% Nicotine extract, 1% Pongamia (14.39%) and 1.5% Pongamia oil (14.05%). The maximum filament length was observed .with 1% nicotine extract + 1% Pongamia oil (417 mt) followed by 1.5% Pongamia oil (396 mt) and 1% Nicotine extract (393 mt). Among the treatments, non-breakable filament length was observed with 2% Pongamia oil (374 mt) followed by 1% Necm oil (371 mt). Denier is the most important cocoon parameter. The lowest Denier was observed with silkworms fed with the leaves sprayed with 1% Nicotine extract(2.1 l)(Table 15.1). No significant difference was observed in other economic characters.

Treated leaves when fed after 14 days of spray, there is no significant difference in mortality (8 - 3%) and other economic

Table 15.1: Residual Effect of Insecticides on Silkworm rearing (7th day of Spray)

Treatment	*Survival (%)*	*SR(%)*	*FL (mt)*	*NBFL (mt)*	*Denier*
Control	89.66	14.52	406	300	2.07
DDVP (0.1%)	93.22	14.02	420	398	2.28
Dimethoate(0.1%)	94.00	14.35	448	448	2.32
1.0% Neem oil	96.00	14.28	417	330	2.18
1.5% Pongamia oil	92.00	13.86	370	371	2.26
2.0% Pongamia oil	96.00	14.66	396	348	2.16
1.0% Nicotine ex.	95.00	15.34	373	337	2.22
2.0% Nicotine ex.	96.66	14.37	393	372	2.36
1.0% Nic.+P. oil(l:l)	90.33	13.8	417	372	2.06
CD at 1%	3.57	N.S.	N.S.	N.S.	N.S.

Table 15.2: Residual Effect of Insecticides on Silkworm rearing (14th day of Spray)

Treatment	*Survival (%)*	*SR(%)*	*FL (mt)*	*NBFL (mt)*	*Denier*
Control	97.00	13.83	301	301	2.00
DDVP (0.1%)	95.00	14.58	304	304	2.07
Dimethoate(0.1%)	95.00	14.3	323	323	2.16
1.0% Neem oil	97.5	14.67	272	272	2.34
1.5% Pongamia oil	93.33	12.66	357	297"	1.87
2.0% Pongamia oil	96.66	15.24	297	247	2.78
1.0% Nicotine ex.	95.00	13.53	364	364	2.2
2.0% Nicotine ex.	97.00	13.67	328	274	2.2S
1.0% Nic. + P. oil(1:1)	92.00	13.65	342	242	2.28
CD at 1%,	N.S.	0.47	65.95	22.34	0.10

Table15.3: Residual Effect of Insecticides on Silkworm rearing (21st day of Spray)

Treatment	*Survival (%)*	*SR(%)*	*FL (mt)*	*NBFL (mt)*	*Denier*
Control	97.30	13.21	314	280	2.23
DDVP (0.1%)	97.00	13.07	365.	266	2.13
Dimethoate (0.1%)	94.00	12.92	298	348	2.05
1.0% Ncemoil	97.60	14.91	299	251	1.99
1.5% Pongamia oil	95.00	12.98	280	234	1.92
2.0% Pongamia oil	97.00	13.20	259	242	2.10
1.0% Nicotine ex.	96.00	13.28	274	244	1.83
2.0% Nicotine ex.	97.30	12.94	318	301	2.10
1.0% Nic. +P. oil(l:l)	94.00	13.35	302	302	2.27
CD at 1%	N.S.	N.S.	N.S.	N.S.	0.20

NBFL - Non breakable filament length; FL - filament length.

characters are on par •with the control, has no significant effect on mortality, survival% are on par with the control and shows no significant differences. The maximum survivability was observed with 1% Neem oil (93%), followed by the combinaiion of 1% Nicotine extract and 1% Pongamia oil (90.66%) etc. The maximum S.R.% was observed with 2% Pongamia oil (15.01%) and followed by 1% Neem oil etc. The maximum filament length was observed in 1% Nicotine extract + 1% Pongamia oil treated leaves (358 mt) followed by 1.5% Pongamia oil (349 mt) and others. The non-breakable filament length was observed to be maximum in 1% Nicotine + 1.0% Pongamia oil treated leaves (358 mt) followed by 2.0% Nicotine extract. The minimum Denier was observed in 1.5% Pongamia oil
1.5% Pongamia oil

(1.87%) followed by 1% Nicotine extract (2.04) and others (Table 15.2). When silkworms were fed for 21 days there was minimum mortality and no significant differences in survival, SR%, filament length and denier among the treatments because of the effect of the botanicals which were degraded with passing of the days, leaving no residue. The observations have revealed that the maximum survivability was observed with 1% Nicotine + 1% Pongamia oil (97.33%), followed by 2.0% Pongamia oil (96.00%) and others. The maximum S.R.% was observed with 1% Neem oil (14.52) followed by 1.5% Pongamia oil (14.09). The maximum filament length was observed with 2.0% Pongamia oil (352 mt) followed by 1.5% Pongamia oil (334 mt). The non-breakable filament length was observed to be maximum with silkworm lots treated with 1.5% Pongamia oil (321 mt), followed by 2% Pongamia oil and 1% Neem oil ((300 mt). The denier was bound to be minimum with the silkworm lots treated with 2% Nicotine extract (1.92) followed by 1.0% Nicotine extract. The result in conformity with studies of Suhas & Deviah (1985) where dichlorvos (0.016%) treated mulberry leaves were safe for feeding to *B.mori* after 20 days, Bandyopadhyay *et al.*, (2002 & 2005) has also opined that 1% Neem oil sprayed leaves when fed after 14 days of spray has minimal effect on the survival (89%) of silkworm *Bombyx mori,* without any impact on economic parameters.

From the present study, it can be inferred that feeding the selected botanicals sprayed leaves to silkworms does not cause any kind of detrimental effect on silkworm, hence they can safely be used in the silkworm rearing after observing 14 days residual/ safe period.

ACKNOWLEDGEMENT

We are grateful to N.B.Kar, Scientist – 'C', Reeling & Spinning section of CSR&TI, Berhampore Institute for his assistance in the reeling analysis of the cocoons.

REFERENCES

Asoka, J. & Patil, B.V. (2001). Bio efficacy of neem based insecticides against mulberry thrips. *In: Proc. of Natl. Sem. on mulberry Res.*, In India, 26-28 Nov. Bangalore, pp 820-824.

Bandyopadhyay, U.K., Kumar, M.V.S., Das, K.K. & Saratcliandra, B.(200l). Yield loss in mulberry due to sucking pest whitefly, *Dialeuropora decempuncta* Quaintance and Baker (Homoptera: Aleyrodidae). *Int. J. Indust. Ento.*, 2(1): 75-78.

Chakraborty, M.K., Prabhu, S.R. & Joshi, B.G. (1976).Isolating toxicants from Pongamia oil to evaluate their insecticidal properties on tobacco caterpillar *(S. litura). Tobacco Research*, 2(l): 38-44,

Mohan,C., Nair, C.P.R. & Rajan, P. (2001). Scope of botanical insecticides in the management of *Oryctes rkinacerous* (Linn.) and *Rhynchophorus ferrugixeus* affecting coconut plant. *Entomon.,* 26:47-51.

David, B.V. & Regupathy, E. (2004). Whiteflies (Homoptera Aleyiodidae) of mulberry, *Moms alba* L., in India. *Pestology,* XXVIII(10): 24-32.

Erier, F., Yegen, O. & Zeller, W. (2007). Field evaluation of botanical products against *Pear Psylla* Homoptera:Psyllidae). *J. Eco. Ento.,* 100(7): 66-71.

Flint, H.M. & Parks, N.I. (1989). Effect of azadirachtin from th cneem tree on mature sweet potato whitefly *B. tabaci* and other selected pest species on cotton. *J. Agri. Ento.,* 6:211-215.

Gupta, G.P. & Lal, R. (1998). Utilization of newer insecticides and neem in cotton pest management. *Ann.Pl. Proc. Sci.*, 6: 155-160.

Gupta, M.P. & Rawat, G..S. (2004). Evaluation of Neem products and their administration with insecticides against bud fly incidence in Linseed. *Ann. of Pl. Pr. Sc.*, 12(1): 1-8.

Heyde, V.D.J, Saxena, R.C. & Schmutterer, H. (1984). Neem oil and and neem extracts as potential insecticides for control of hemiltcrous rice pests in natural pesticides from the neem tree and other tropical plants. *Proc., 2nd Int. Neem Conf.* pp. 377-390.

Murray, B.I. (1999). Neem and related natural products. In: *Biopestcides Use and Delivery* (Eds. Mall &Menn). pp. 139-154.

Joshi, B.G. & Seetharamaiah (1975). Note on relative toxicity to some *Nicotinia* sp. to tobacco caterpillar, S. litura Fab.. *Tobacco Res.*, 4(1): 86-87.

Mann, G.S., Dlialliwal, G.S. & Dhawan, A.K. (2001). Effect on alternative application of Neem products and insecticides on population of *B. labaci* and its impact on ball worm damage in upland cotton. *Ann. of Pl. Proct.*, 9(1): 22-25.

Mukhopadhyay, S.K., Bandyopadhyay, U.K. & Kumar, M.V.S. (2005). Whitefly infestatation on mulberry, its integrated management and forecasting. In *proc. of congress of International Sericultural Commission* held at Bangalore on 15-18 Dec., 2005, pp. 79-81.

Mukhopadhyay, S. K., Santha Kumar, M.V. & Das, S.K. (2006). Management of thrips *Pseudodendrothrips man* (niwa) in mulberry through botanicals. In *Proc. workshop Appropriate technologies for mulberry sericulture in eastern and north eastern India* held at Berhampore on 17-18 January, 2006.

Mukhopadhyay, S. K. (2006). Technologies for management of mulberry In *Proc. Workshop Appropriate technologies for mulberry sericulture in eastern and north eastern India* held at Berhampore on 17-18 January, 2006, p. 14.

Narayanaswamy, K.C., Ramegnwda, T., Raghuraman, R. & Manjunath, M.S. (1999). Biochemical changes in spiraling whitcfly (*A. dispersus*) infested mulberry leaves and their influence on some economic parameters of silkworm. *Entomon.*, 24: 215-220.

Nath, P., Bhusan, S. & Singh, A.K. (2002). Evaluation of Neem based formulations and Neem seed kernels extract against aginst the insect pest of sesame. *Annl. Pl. Proct. Sc.*, 10: 207-211

Nimbalkar, S.A., Khodke, S.M., Taley, Y.N. & Patil, K.J. (1990). Bio efficacy of Neem and Karanj seed extracts along with commercial plant products against ball worm on cotton. In *Botanical pesticides in IPM, ISTS*, Rajmundry, pp. 252-55.

Patel, C.V., Clhala, R., Patel, N.M. & Saha, H. (1990). Field efficacy of Neemark in comparison to chemical insecticides against Sugarcane whitefly, *A. barulensis. Proceedings of symposium on Botanicals in IPM.* pp. 343-350.

Pillai. K. & Purnea, S. (1988). Neem for control of Rice thrips. *International Rice Research News Letter*, 13:33-35

Pitarelli, G.W., Bute, J.G., Neil, J.W., Lusby, W.R. & Waster, R.M. (1993). Biological pesticides derived from *Nicotinia* plants. US Patent No. 5260281.

Prabhu, S.R., Chakraborty, M.K., Sitaramarih, S., & Joshi, B.G. (1981). Isolation and identification of toxic materials from *N. gossei. Tob. Sci.*, 25: 114-7.

Ray, D.E. (1991). Pesticides derived from plants and other organism. In *Handbook of pesticide & toxicology*. (Eds. Hayes, W. J. Jr., Laws, E. R. Jr.). Academic Press, NY.

Shah, M.A.S., Singh, C.H. & Varatharajan, R. (2005). Effect of Neemazol on onion thrips. *Thrips tabaci. Ann. Pl. Protect. Sc.*, 13(2): 470-71.

Sundararaj, R.S., Mureresan, R.N. & Misra (1995). Efficacy of Neem seed oil against Babool whitefly, *A. rachipora. Indain Forester*, 121: 1077-79.

Sharma, R.N., Tare, V., Pawar, R., Pawar, S.S. & Patwabardhan, S. S. (1999). Potentials of some plants for development of new and safer pest control agents of selected pests of agn. and horticulture. In *Green Pesticides Crop Protection of and Safety evaluation*. Society of Pesticides Sc. India, pp. 104-14.

James, F.W. (1998). Commercial experience with Neem products. In: *Biopestcides Use and Delivery* (Eds. Hall & Menn). pp. 155-170.

16

BENEFICIAL EFFECTS OF GIBBERELLIC ACID (GA_3) ON GROWTH AND REPRODUCTIVE EFFICIENCY OF SILKWORM, *BOMBYX MORI* L.

*C.R. Hegde, S.R. Ananthanarayana and T.M. Veeraiah**

ABSTRACT

Exogenous application of many compounds like moulting hormone, juvenile hormone, anti-juvenile hormone, growth promoters and feed supplements have influenced the growth and development of silkworm, *Bombyx mori* L. In this context, a study was conducted by administering *Gibberellic acid* (GA_3), a well known growth regulator to 4th and 5th instar larvae of parent breeds of ruling multibivoltine hybrid. Three doses of GA_3 viz., 50ppm, 100ppm, and 150ppm were administered to the larvae every day from 1st day of 4th instar till spinning in different frequencies. The frequency of administration was either once, twice or three times in a day. Effect of GA_3 administration on rearing parameters. cocoon parameters and egg production parameters were recorded. The results revealed that administration of 50ppm of GA_3 to Pure Mysore larvae two times every day has improved the weight of matured larvae, cocoon weight, shell weight, and also egg production parameters, where as in CSR2 breed these parameters were improved by 100ppm. Administration of 150ppm has adversely affected all these traits. Ripening of larvae was preponed in all the treated batches as compared to control. The beneficial effects of *Gibberellic acid* in silkworm rearing are discussed in detail.

Key words: Gibberellic acid, growth promotion, larval ripening, reproduction.

Department of Sericulture, Jnanabharathi Campus, Bangalore University, Bangalore-56

*NSSO Head Quarters, Central Silk Board, Bangalore-68.

E-mail: yahoo326@yahoo.com

INTRODUCTION

The mulberry silkworm, *Bombyx mori* L. feeds solely on the foliage of *Morus alba* throughout its lifecycle. However, extensive studies have been made on the exogenous administration of many biologically active substances like juvenile hormone, moulting hormone, anti-juvenile hormone, growth promoters and feed supplements. (Akai and Kobayashi, 1971, Hasegawa 1972, Aomori *et al.*, 1977, Bhaskar 1983, Kiguchi 1984, Thygaraja 1985, Chowdhary *et al.*, 1996, Magadum and Hooli, 1989, Shivakumar *et al.*, 1995, Ninagi and Maruyama, 1996, Vanishree *et al.*, 1996, Chandrakala *et al.*, 1998, Rajegowda 2002)

Gibberellic acid is found to occur naturally in isoprenoid pathway in several living organisms, which generates many metabolically important compounds, since the intermediates in the pathway are the same for these organisms. There is an impact of mutual interaction through the food chain. Synthetic *Gibberellic acid* is available commercially and is cost effective. According to Fletcher and Osborne (1965) smearing of GA_3 on the harvested leaves retards the senescence and increases protein level in many plant species including mulberry. Due to the intricate relationship between the plant, the insect and the growth regulator, it is interesting to study the effect on the growth and metamorphosis of the silkworm.

MATERIALS & METHODS

The larvae of parent breeds of ruling multibivoltine hybrid, namely *Pure Mysore* and CSR_2 were used as experimental animals for this study. Mass rearing of larvae of these breeds was carried out from brushing to end of the 3rd moult by adopting standard procedure as suggested by Krishnaswamy *et al.* (1978). From 1st day of 4th instar the larvae were grouped into 10 batches and fed different doses of the hormone till completion of the larval period. First 3 batches received 50ppm of GA_3, 2nd 3 batches received 100ppm and next 3 batches 150ppm. Tenth batch that did not receive any hormone, served as control. Gibberellic acid was administered to 4th and 5th instar larvae either once, twice or 3 times in a day after smearing it to the mulberry leaves. The mulberry leaves were cut into 2" x 2" size pieces and 20ml of hormone solution was sprayed to each kilogram of leaf pieces in the form of fine particles using a hand sprayer. The sprayed leaf pieces were thoroughly stirred in a plastic container to ensure uniform smearing of the hormone solution. Then the leaf pieces were allowed for air drying for about 20-30min and fed to the worms. V_1 variety mulberry maintained at Jnanabharathi campus of Bangalore University, was utilized as the source of food for the rearing. The details of the treatments are as mentioned below.

T_1 - 50ppm once + remaining 3 normal feedings
T_2 - 50ppm twice + remaining 2 normal feeding
T_3 - 50ppm thrice + remaining 1 normal feeding
T_4 - 100ppm once + remaining 3 normal feeding
T_5 - 100ppm twice + remaining 2 normal feeding
T_6 - 100ppm thrice + remaining 1 normal feeding
T_7 - 150ppm once + remaining 3 normal feeding
T_8 - 150ppm twice + remaining 2 normal feeding
T_9 - 150ppm thrice + remaining 1 normal feeding
Control – No hormone administration.(All normal feedings)

Each treatment was replicated 3 times with 100 larvae per replication. To assess the impact of the hormone administration, important rearing parameters like weight of grown up larvae, Effective rate of rearing (ERR), cocoon yield, single cocoon weight, shell weight, pupation rate and commercial egg production parameters were studied. The data was statistically analysed as per DMRT (Gomez and Gomez,1964).

RESULTS AND DISCUSSION

Results revealed significant variations in almost all the parameters studied in both the races (Fig. 16.1-Fig. 16.13).

Body weight of matured Larvae

Gibberellic acid administration has resulted in significant improvement in larval growth in majority of the treated batches. Administration of 100ppm GA_3 two times every day to the CSR2 larvae has significantly increased the weight of 10 matured larvae (41.335g) over the control & other dosages. Where as in *Pure Mysore* maximum larval weight was obtained (31.15g) with a lower dose of 50ppm, at the same frequency of administration (Fig.16.1). As the dosage increased to 150ppm there was an adverse effect on this parameter in both the races. This indicates that GA_3 has adverse effect at higher dosage. These results agree with the results of Magadum *et. al.,* (1990) who obtained increase in larval weight by application of methoprene during II to V stage larvae.

Effective Rate of Rearing (ERR)

Administration of 100ppm GA_3 twice a day had a significant positive impact on the ERR both by number(9517) and by weight(17.499kg) in CSR2. In *Pure Mysore,* 50ppm application at the same frequency resulted in higher ERR by weight (11.86kg) but not the number. This indicates that *Pure Myore* larvae are not influenced by GA_3 with regard to survivability (Fig. 16.2 & 16.3).

Cocoon Characters

The cocoon weight and shell weight are maximum(1.858g & 0.42g) when CSR2 larvae were administered with 100ppm solution of the hormone two times a day. But the shell ratio remained unaffected by the hormone administration. In *Pure Mysore,* similar trend was observed in 50ppm batch with same frequency of administration (Fig.16.4-16.6).

Pupation Rate

Pupation rate has significantly increased in CSR2 when 100ppm of the hormone solution was administered two times a day(91.67%) where as in *Pure Mysore* the treatment of GA_3 did not have any significant effect on this parameter(Fig.16.7).

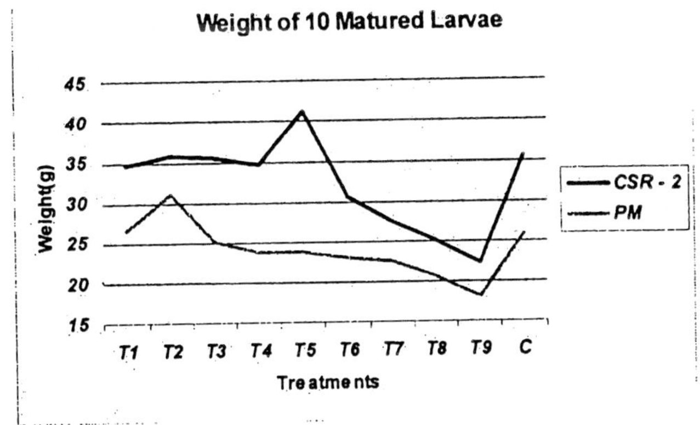

Fig. 16.1

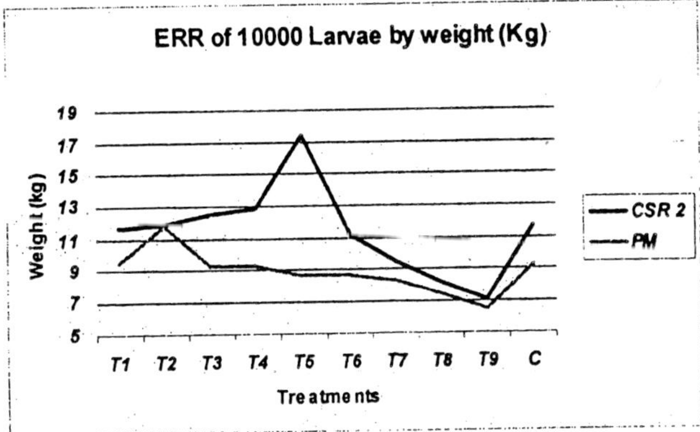

Fig. 16.2

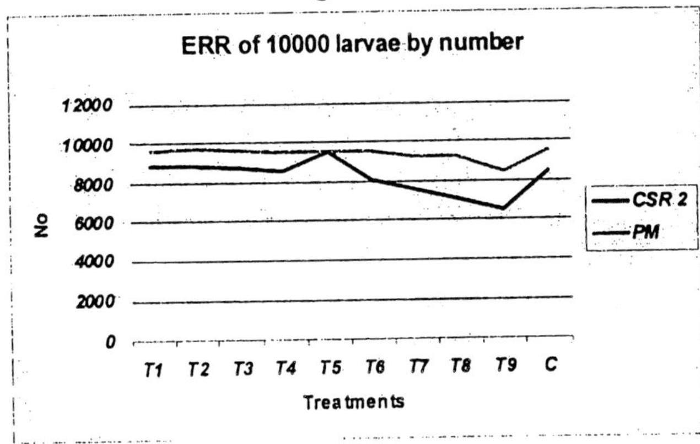

Fig. 16.3

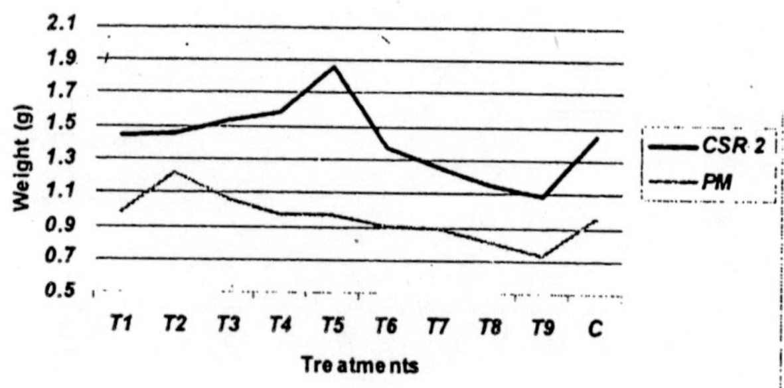

Fig. 16.4

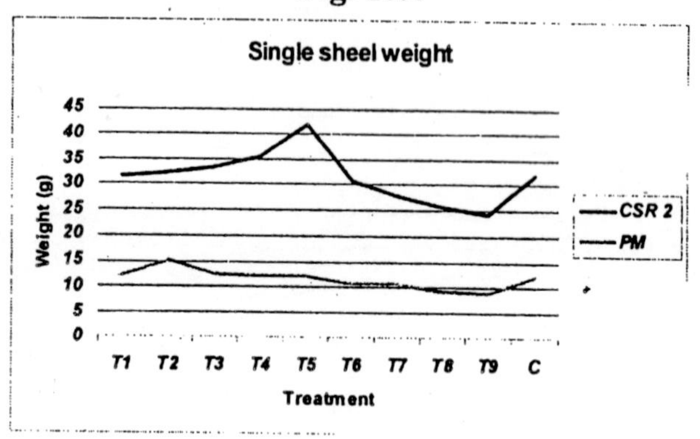

Fig. 16.5

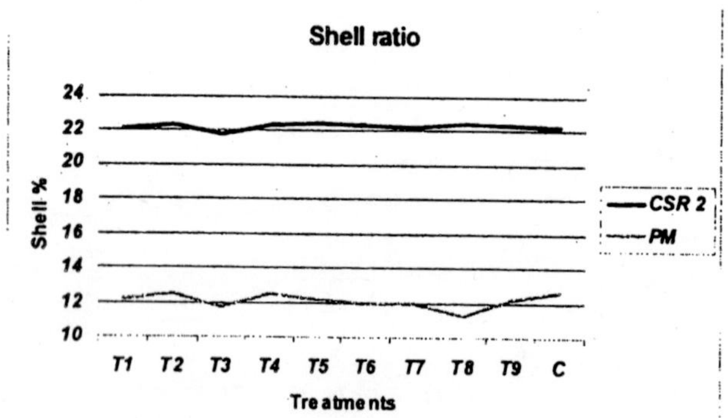

Fig. 16.6

Adult moth Eclosion

The rate of moth eclosion is an important parameter for a commercial silkworm egg producer. In CSR2 both male and female adults are utilized for seed production where as in *Pure Mysore,* only females are used. As the pupation rate was higher in 100ppm treated batches of CSR2 the moth emergence ratio of

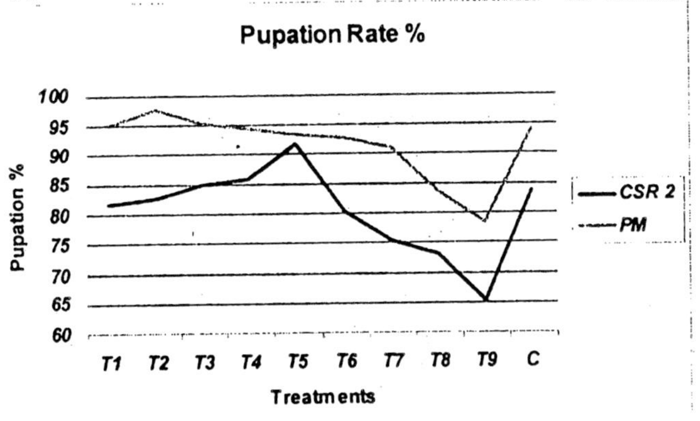

Fig. 16.7

both male and females were also higher (91.33& 68.49%) in this treatment when compared to control & other treatments. In *Pure Mysore,* female moth emergence was maximum when the hormone was applied two times at the concentration of 50ppm.(Fig.16.8 & 16.9).

Reproductive Parameters

Reproductive parameters such as layings recovery ratio, weight of single egg and the unfertilized egg ratio also significantly varied in treated batches. Fecundity was not affected by any of the treatments. Laying recovery significantly improved both in CSR2 and *Pure Mysore* breeds in 100 ppm and 50 ppm GA3 administered batches, respectively (36.33 &41.67%)(Fig.16.10). Weight of individual egg also showed similar trend in both the races(Fig.16.12). Unfertilized egg occurrence was minimum in CSR2 moths, which received 100ppm GA3 in their larval stage. In *Pure Mysore*

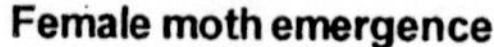

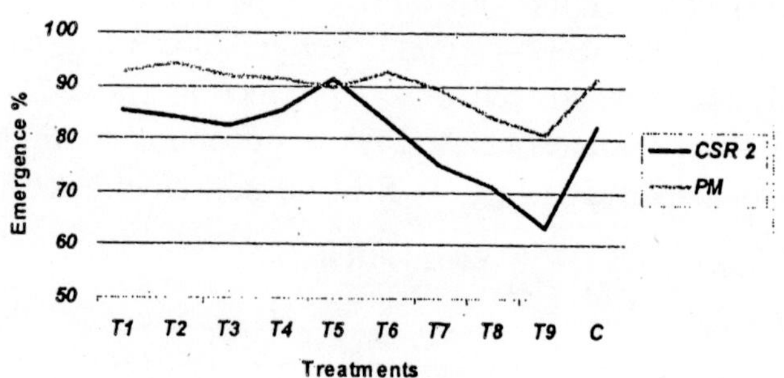

Fig. 16.8

Malemoth emergence ratio

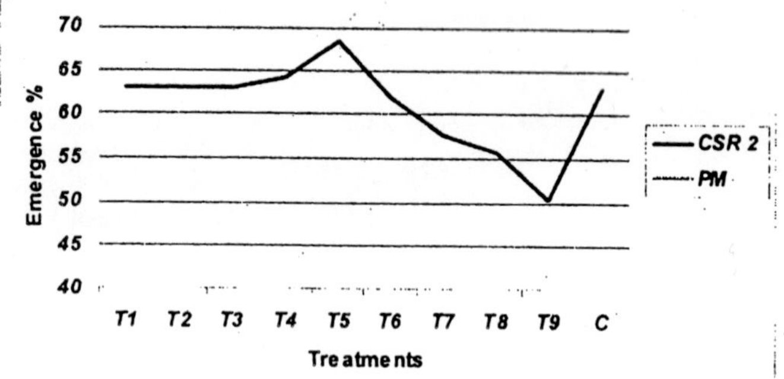

Fig. 16.9

Unfertilized egg occurrence reduced significantly in 50ppm batch (Fig. 16.13).

Considerable number of studies have been conducted before with various Juvenile hormones and Moulting hormones. The results of the present study are in agreement with Mugadum *et al.*, (1990) who obtained significant increase in larval weight and cocoon weight by administering 2.5ppm methoprene to II to V instar larvae. Mugadum and Mugadum(1996) reported that topical application of methoprene to *Pure Mysore* larvae extended the larval duration. On the contrary, in the present study application of Gibberellic acid reduced the larval period by 18-24h. By

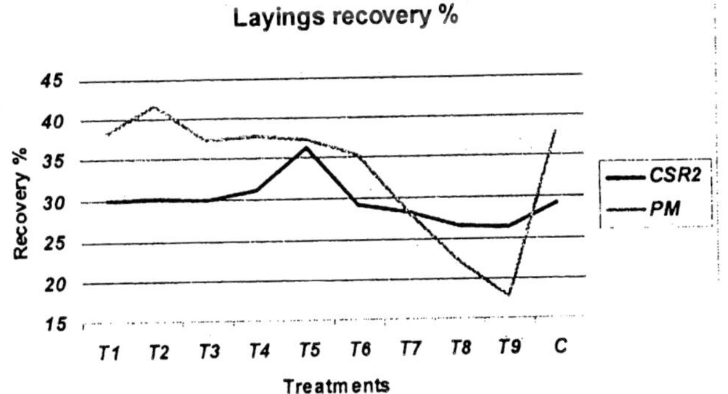

Fig. 16.10

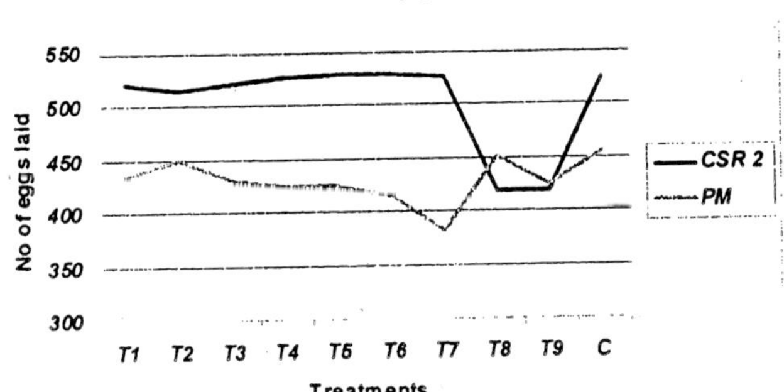

Fig. 16.11

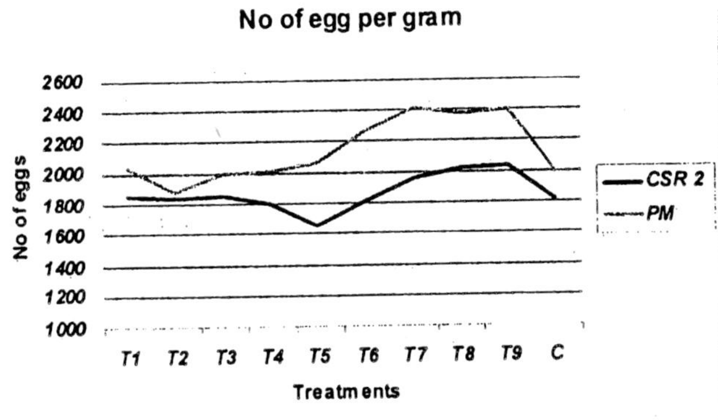

Fig. 16.12

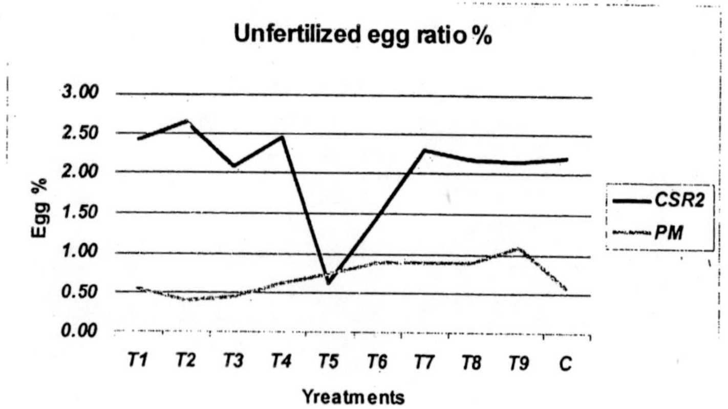

Fig. 16.13

application of methoprene Zhuang *et al.* (1992) were able to increase the cocoon yield and out come of present study agrees with this result.

The present study has been supported by the earlier report that R394, a juvenile hormone improves the shell weight without prolonging the larval period. The present study is also supported by the results of Santhakumari *et al.* (1989), where in they obtained higher larval weight; cocoon weight and cocoon yield by administration of 100ppm GA_3 to multibivoltine hybrid larvae. They have also reported improvement in fecundity and egg weight.

The improved effects of the GA_3 on the silkworm could be mediated either by direct effect of the plant hormone on the metamorphosis of the insect or the indirectly by its influence on the mulberry leaves. The conclusion of this study is that application of GA_3 to detached mulberry leaves as a supplement to the silkworm diet is beneficial for the sericulture industry.

ACKNOWLEDGEMENTS

The authors thankfully acknowledge Mr.Nanjunda Shastri Statistician, CSTRI, Bangalore for extending guidance and help in drawing statistical analysis. Authors also thank Dr. B.V. Narahari

Rao, Scientist (Retd) KSSRDI, Bangalore for encouragement and help in carrying out this work.

REFERENCES

Akai, H. & Kobayashi, M. (1971). Induction of prolonged larval instar by the juvenile hormone in *B. mori* L (Lepidoptera: Bombycidae). *Appl. Entomol. Zool.*, 6: 1938-1939.

Aomori, S., Ozawa, Y. & Nihmura, M. (1977). Timely administration of synthetic compounds with the juvenile hormone activity to silkworm larvae reared on an artificial diet. *J. Seri. Sci. Jpn.*, 46(1): 69-76.

Bhaskar, M., Bharathi, D., Reddanna, P. & Govindappa, S. (1983) Growth pattern of Silkworm, *Bombyx mori* larvae on exposure to Prolactin. *Indian J. Seric.*, 22: 33-35.

Chandrakala, M.V., Maribashetty, V.G. & Jyothi, H.K., (1998). Application of phytoecdy - steroids in sericulture. *Curr. Sci.*, 74(4): 341-346.

Chowdhary, S.K., Khajuria, R. K., Kulsrasta, A.K, Ogra, R.K. & Jain, S.M. (1996). Silk enhancer from *Cassia tora* for *B. mori* L. *Sericologia,* 36(2): 259-266.

Fletcher, R.A. & Osborne, D.J. (1965). Regulation of protein and nucleic acid synthesis by gibberellins during leaf senescence. *Nature,* 207: 1176-1177.

Gomez, K.A. & Gomez, A.A. (1984). *Statistical Procedures for Agricultural Research.* John Willey and Sons, pp. 644-645.

Hasegawa, K. & Ahmed, M.A. (1972). Penetration of phytoecdysones through the pupal cuticle of the silkworm, *B. mori. J. Insect Physiol.,* 18: 959-971.

Kiguchi, K., Mori, T. & Akai, H. (1984). Effects of anti-juvenile hormone 'ETB' on the development and metamorphosis of the silkworm, *Bombyx mori. J. Insect Physiol.*, 30(6): 499-506.

Krishnaswamy, S. (1978). New Technology of Silkworm Rearing, Bulletin No. 2 CSRTI, Mysore, pp. 1-24.

Mugadum, S.B. & Hooli, M.A. (1989). Effect of insulin on the polyvoltine silkworm, the Pure Mysore breed of *Bombyx mori* L. *Sericologia,* 29(1): 17-27.

Mugadum, S.B., Hooli, M.A. & Mugadum, V.B. (1990). Effect of Methoprene on larval and cocoon weight, ovariole length, egg number and fecundity of *Bombyx mori* L. *Entomol. Hellenica*, 8: 37-40.

Magadum, V.B. & Magadum, S.B. (1996. Effect of ZR 515 on the Larval period, silkgland weight, cocooning and moth emergence in polyvoltine silkworms, *Bombyx mori* L. *PKV Res. J.*, 20(1): 21-24.

Ninagi, O. & Maruyama, M. (1996). Utilization of 20-Hydroxyecdysone extracted from a plant, in sericulture. *JARQ*, 30: 123-128.

Rajegowda (2002). Impact of 'Seripro' on cocoon production and productivity in silkworm, *Bombyx mori* L. *Advances in Indian Sericulture Research. Proceedings of the National conference on Strategies for Sericulture research and Development.* 16-18, Nov. 2000, CSR&TI, Mysore.

Santakumari, M., Shreedhar, V.M. & Fletcher, R.A. (1989). Gibberellic acid improves the commercial characteristics of silkworms. *Sericologia,* 29(3):347-351.

Shivakumar, G.R., Raman, K.V.A., Magadum, S.B., Datta, R.K., Hussain, S.S., Banerji, A. & Chowdhary, S.K.(1995). Effect of Phytoecdysteroids on larval maturaion and economic parameters of the silkworm, *Bombyx mori* L. *Indian J. Seric.*, 34(1): 46-49.

Thygaraja, B.S., Jolly, M.S., Datta, R.K. & Murthy, C.V.N. (1985). Effect of thyroxine on silk improvement in *Bombyx mori* L. *Indian J. Seric.*, 24(2): 77-82.

Vanishree, V., Nirmala, X. & Krishna, M. (1996). Response of five different races of silkworm, *Bombyx mori* (Lepidoptera:Bombycidae) to hydrolysed soya protein supplementation. *Sericologia*, 36(4):691-698.

Zhuang, D.H., Xiang, M. & Gui, Z.Z. (1992). The practical studies on the Growth and development by insect hormones in the silkworm, *Bombyx mori*. *XIX Int. Cong. Entomol.*, Jun. 28 - Jul. 2, 1992, Beijing, China.

SEASON SPECIFIC SILKWORM REARING PACKAGE FOR EASTERN INDIA

T. Datta (Biswas), A.K. Saha, S.K. Das and A.K. Bajpai

ABSTRACT

Eastern India is characterized by luxuriant growth of mulberry due to high rainfall and fertile soil. But varied climatic condition of this region with extremely hot and humid as well as hot and dry summer spells and severe winter with very low humidity is a major impediment for silkworm rearing, for which silkworm rearers experience frequent crop loss. Present package of practices of silkworm rearing is not enough to reach to the desired level of cocoon production even with the use of productive silkworm breeds/hybrids. To combat the situation, manipulation of major abiotic factors like temperature and relative humidity is needed to make them congenial for silkworm rearing by developing season specific rearing package of practices for silkworm. Keeping this in view, an attempt was made at CSR & TI, Berhampore consecutively for three years (2003-2005) for developing a season specific silkworm rearing package for Eastern India to increase cocoon productivity and quality for agro climatic condition of this region. By utilizing this package of practices, a farmer can earn an additional cocoon yield of 9-10 kg and 11-12 kg 7100 dfls during favourable and unfavourable crop seasons respectively which will fetch an additional income of Rs. 900.00 - Rs.1000.00(cocoon rate Rs. 100/kg) and Rs. 880.00 - Rs. 960.00 (cocoon rate Rs. 80/kg) respectively.

Key word: Silkworm; Eastern India; Standard Package; Humidity; Chauki Rearing.

Central Sericultural Research & Training Institute (CSR&TI), Central Silk Board, Berhampore-742101, W.B.

E-Mails: tdattabiswas@rediffmail.com; akbjp@rediffmail.com; sahaatul@rediffmail.com

INTRODUCTION

Eastern India is characterized with widely fluctuating climatic condition with extremely hot and humid as well as hot and dry summer and severe winter with very low humidity. During summer, the temperature reaches beyond 45°C with low humidity ranging between 35 & 50% while in rainy season the temperature some time reaches 38°C accompanied with about 90% relative humidity which are not at all congenial for silkworm rearing for which silkworm rearers frequently experience crop loss, sometimes total crop failure.

The commercial crop seasons are broadly classified into favourable (when climatic condition is more or less congenial for silkworm rearing) and unfavourable (when climatic condition is severe with high temperature and high humidity) depending upon the climatic condition.

It is well known that the growth and development of the silkworm is greatly influenced by microclimatic conditions such as rearing bed humidity and temperature. Therefore, the rearing management assumes special importance in order to create the optimum environment for the larval growth (Rajan *et al.*, 1995).

Keeping this in view, the present study was under taken for developing a season specific rearing package suitable for Eastern India for qualitative and quantitative improvement of cocoon production inspite of varied climatic condition. In Eastern India bamboo spiral mountage, popularly known as Chandraki, is commonly used for spinning of silkworm . Considering the advantages of storing and mounting space, cost and cocoon quality, which are the main concern for most of the marginal sericulture farmers of Eastern India, feasibility of using plastic collapsible mountage in this agro-climatic zone along with the optimum larval density was also studied for incorporation in the recommended package.

MATERIALS & METHODS

Three years study (2003-2005) was undertaken at the Silkworm Physiology & Rearing Technology Innovation Laboratory of Central Sericultural Research & Training Institute , Berhampore, West Bengal, during both favourable (November, February, April) and unfavourable (June-July and August-September) commercial crop seasons.

For rearing, S1635 variety of mulberry was used in all the seasons to avoid any nutritional effect. Standard package of practices were followed for mulberry cultivation.

FOR CHAWKI REARING

Three methods of chawki rearing viz. Box rearing (trays with bottom and top paraffin paper surrounded by wet foam pad in piled condition), Wrap up method (rearing bed in the trays wrapped up with paraffin paper on all the four side), Open type of shelf rearing (individual tray with only bottom paraffin paper) along with two type of bed spacing (3.75 - 45 sq.ft../100 dfls, earlier recommendation for Southern India and 4.50 - 54 sq.ft./100 dfls) were tested keeping farmers practice (shelf rearing with 3-35 sq.ft./ 100 dfls bed spacing) as control.

EXPERIMENTS

T1 = Open Type Shelf Rearing + 3.75 - 45 sq.ft../100 dfls bed spacing

T2 = Open Type Shelf Rearing + 4.50 - 54 sq.ft../100 dfls bed spacing

T3 = Box Rearing + 3.75 - 45 sq.ft../100 dfls bed spacing

T4 = Box Rearing + 4.50 - 54 sq.ft../100 dfls bed spacing

T5 = Wrap up Rearing + 3.75 - 45 sq.ft../100 dfls bed spacing

T6 = Wrap up Rearing + 4.50 - 54 sq.ft../100 dfls bed spacing

•Control = Shelf Rearing + 3.00 - 35 sq.ft../100 dfls bed spacing

* Farmers' practice

FOR LATE AGE REARING

Three times and four times feeding per day at an interval of 8 hours and 6 hours respectively, 9" gap between the rearing trays and bed spacing of 360 sq.ft../100 dfls (earlier recommendation for Southern India) & 396 sq.ft. per 100 dfls were undertaken keeping 3 feeding per day at an interval of 8 hours, 6" gap between the rearing trays and 288 sq.ft bed spacing per 100 dfls as control (Farmers' practice).

EXPERIMENTAL DETAILS

T1 Shelf rearing + 9" gap between the + 3 feeding + 360 sq.ft.bed spacing /ays 100 dfls

T2 Shelf rearing + 9" gap between the+ 4 feeding + 360 sq.ft. bed spacing/trays 100 dfls

T3 Shelf rearing + 9" gap between the+ 3 feeding + 396 sq.ft. bed spacing/trays 100 dfls

T4 Shelf rearing + 9" gap between the+ 4 feeding + 396 sq.ft. bed spacing/trays 100 dfls

*Con. Shelf rearing + 6" gap between the + 3 feeding + 288 sq.ft bed spacing/trays 100 dfls

* Farmers' practice

Disinfection of rearing room, incubation, blackboxing, bed cleaning, bed disinfection, moulting care, maintenance of environmental condition etc. were done as per the standard schedule.

FOR MOUNTING

Two types of mountages i.e. traditional bamboo spiral mountage and plastic collapsible mountage were tested. Variation in larval density as shown in the experiment details was also tested for both types of mountages. For bamboo spiral mountage spinning in both indoor and outdoor condition was considered. In seriposition the worms were mounted differently according to the treatments.

TREATMENT DETAILS

T1	Bamboo spiral mountage	+ 50*/60** larvae/sq.ft.+Outdoor spinning.
T2	Bamboo spiral mountage	+ 60*/72** larvae/sq.ft + Outdoor spinning.
T3	Bamboo spiral mountage	+ 50*/60** larvae/sq.ft + Indoor spinning.
T4	Bamboo spiral mountage	+ 60*/72** larvae/sq.ft + Indoor spinning.
T5	Plastic	collapsible + 50*/60** larvae/sq.ft + Indoor spinning mountage
T6	Plastic	collapsible + 60*/72** larvae/sq.ft + Indoor spinning mountage
#Con	Bamboo spiral mountage	+ 65*/80** larvae/sq.ft + Spinning in open air

Farmers' practice, * multi x bi, ** multi x multi

COMBINATION USED

Favourable crop season : multi x bi (N x YB) hybrid

Unfavourable crop seasons: multi x multi (N x M12W) hybrid

PARAMETERS RECORDED

Yield 710,000 larvae (No.& Wt. in kg) shell%, filament length, renditta, and reelability% were recorded for chawki and late age rearing. For mounting, cocooning% and Floss% were recorded.

During rearing each treatment was replicated three times with 500 larvae per replication after 3rd moult. Data were collected for three consecutive years in all the 5 commercial crop season of West Bengal. Three years pooled data were statistically analyzed separately for favourable seasons (Nov., Feb. & April) and unfavourable seasons (June-July & Aug.-Sept.) to get the season specific recommendation.

RESULT

A. FOR CHAWKI REARING

A. Favourable Seasons

Volt has been found from Table 17.1 that during favourable season Box rearing with 4.5 -54 sq.ft spacing per 100 dfls performed best for most of the rearing and reeling parameters

viz. yield/10,000 larvae (no. & wt.), Filament Length, Renditta and reelability% though Shell% showed at par value.

Box rearing Fig. 17.1 and Wrap-up method of chawki rearing with both 3.75 - 45 sq.ft. and 4.50 - 54 sq.ft. bed spacing/100 dfls were found significantly superior than farmers' present practice i.e. control which may be due to the fact that both methods provide more or less optimum micro-climatic condition for optimum growth and development of larvae leading to improvement of economic characters

B. Unfavourable Season:

Analyzed data (Table 17.2) revealed the superiority of open type of shelf rearing with 4.50 - 54 sq.ft. bed spacing/100 dfls with respect of most of the rearing and reeling parameters. Box rearing, Wrap-up method of chawki rearing and Open type of shelf rearing with both type of spacing (3.75 - 45 sq.ft. & 4.50 - 54 sq.ft) performed significantly better in rearing and reeling parameters in comparison to present farmers' practice (control).

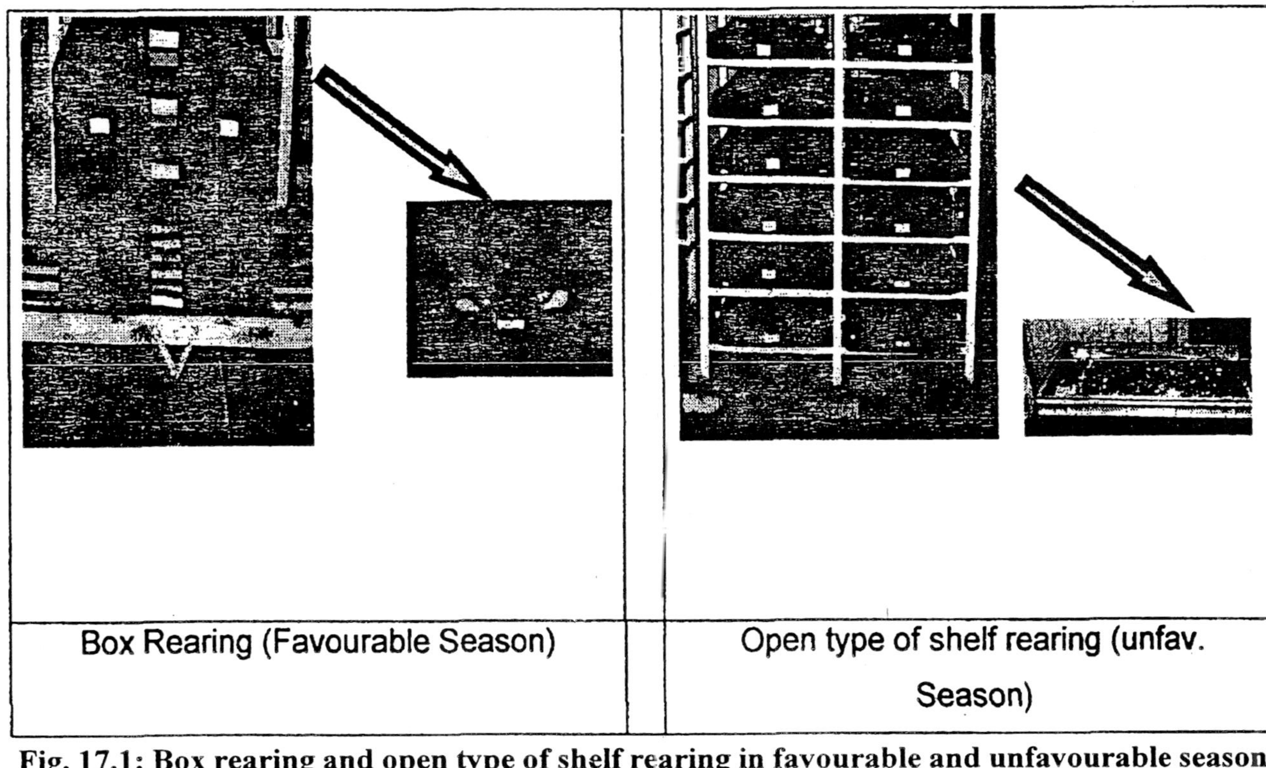

Fig. 17.1: Box rearing and open type of shelf rearing in favourable and unfavourable seasons respectively

Table 17.1 : Rearing and reeling performance with recommended and traditional method of chawki rearing during favourable season.

Treatment	Yield/ 10000 Larvae		Shell %	Fl.(m)	Renditta	Reelability %
	No.	Wt. (Kg.)				
T1	7272	10.93	16.28	654	8.66	76.55
T2	7389	10.98	16.10	660	8.43	76.30
T3	8077	12.62	**17.22**	705	8.31	78.39
T4	**8181**	**12.98**	**17.22**	**718**	**8.07**	**79.69**
T5	7937	12.03	16.95	691	8.37	79.52
T6	7943	12.01	16.71	689	8.37	78.44
Control	7269	10.50	16.23	646	8.74	75.90
CD at 5%	**142.9**	**0.31**	**0.26**	**13.2**	**0.13**	**1.13**
CV%	**3.408**	**4.85**	**2.877**	**3.61**	**2.907**	**2.70**

Table 17.2 : Rearing and reeling performance with recommended and traditional method of chawki rearing during favourable season.

Treatment	Yield/ 10000 Larvae		Shell %	Fl.(m)	Renditta	Reelability %
	No.	Wt. (Kg.)				
T1	8695	10.25	14.36	419	12.01	66.61
T2	**8860**	**10.44**	**14.48**	**427**	11.11	**71.07**
T3	8219	9.64	13.96	413	**11.39**	67.19
T4	8362	9.79	14.14	425	11.61	68.51
T5	8688	10.17	14.33	418	11.93	66.64
T6	8569	10.11	14.41	423	12.00	67.21
Control	7005	7.68	14.07	385	12.08	64.58
CD at 5%	**133.72**	**0.19**	**0.17**	**13.24**	**0.2**	**1.68**
CV%	**2.909**	**3.461**	**2.241**	**5.881**	**3.124**	**4.594**

RECOMMENDED METHOD OF CHAWKI REARING

B. FOR LATE AGE REARING:

A. Favourable Season:

It may be observed from Table 17.3 that during favourable season four feeding per day with art interval of 6 hours with 396 sq.ft. bed spacing/100 dfls and 9" gap between the trays

with all other standard methods performed best for most of the rearing and reeling parameters like yield per 10,000 larvae, shell% , filament length, renditta and reelability%.

B. Unfavourable Seasons :

Analyzed data (Table 17.4) revealed the superiority of three feeding per day with an interval of 8 hours with 396 sq.ft spacing/ 100 dfls and 9" gap between the trays with respect to most of the parameters studied but shell% showed non significant variation between the treatments.

C. FOR MOUNTING

A. Favourable Seasons

It is clear from Table 17.5 that in all the three favourable seasons collapsible mountages with 50/60 (multi x bi & multi x multi) larvae per sq.ft. performed best for most of the reeling parameters. Reelaiblity%, the most important post cocoon parameter, was found significantly higher in T5 than T6 in all the three seasons. Farmers' practice i.e. control was found inferior to all the treatments.

B. Unfavourable Seasons

During unfavourable season also T5 showed its superiority (Table 17.6) in respect of most of the parameters. Though cocooning% was some what lower in collapsible mountage due to urination problem, it was found statistically non significant.

Table 17.3 : Rearing and reeling performance with recommended and traditional method of late age rearing during favourable season.

Treatment	Yield/10000 larvae		Shell %	Fl.(m)	Renditta	Reelability,%
	No	Wt.(Kg.)				
T1	7457	10.79	16.29	651	8.68	76.19
T2	8326	12.06	16.84	707	8.37	78.43
T3	7512	10.79	16.41	648	8.50	75.37
T4	**8481**	**12.48**	**16.98**	**717**	**8.22**	**79.24**
Control	7222	10.24	16.12	648	8.74	75.67
CD at 5%	**113.08**	**0.24**	**0.19**	**12.87**	**0.12**	**1.44**
CV%	**3.611**	**5.747**	**2.2735**	**4.049**	**2.432**	**3.91**

Table 17.4 : Rearing and reeling performance with recommended and traditional method of late age rearing during unfavourable season.

Treatment	Yield/10000 larvae		Shell %	Fl.(m)	Renditta	Reelability %
	No	Wt. (kg.)				
T1	8553	9.91	14.26	425	11.52	**68.81**
T2	8206	9.39	14.24	420	11.46	67.74
T3	**8659**	**10.03**	**14.32**	**445**	11.50	67.94
T4	8438	9.74	14.06	436	**11.10**	68.63
Control	8347	9.59	14.03	396	11.78	64.06
CD at 5%	**215.44**	**0.271**	**NS**	**12.471**	**0.24**	**2.027**
CV%	**4.69**	**5.118**	**2.548**	**5.398**	**3.841**	**5.521**

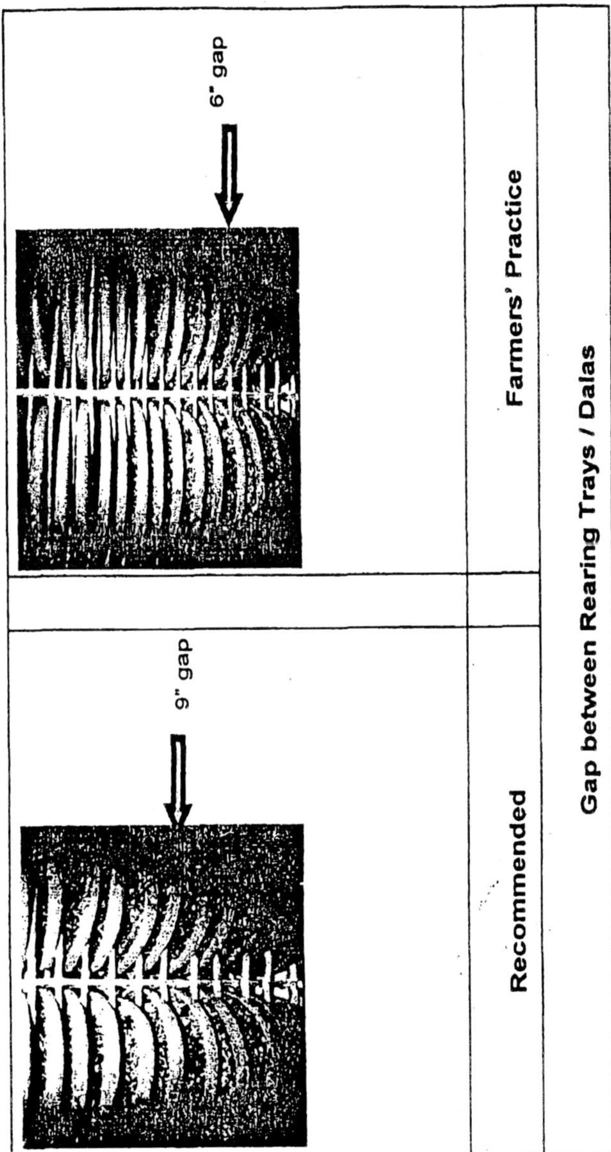

Fig. 17.2: Gap (recommended and farmer's practice) between rearing trays/dalas

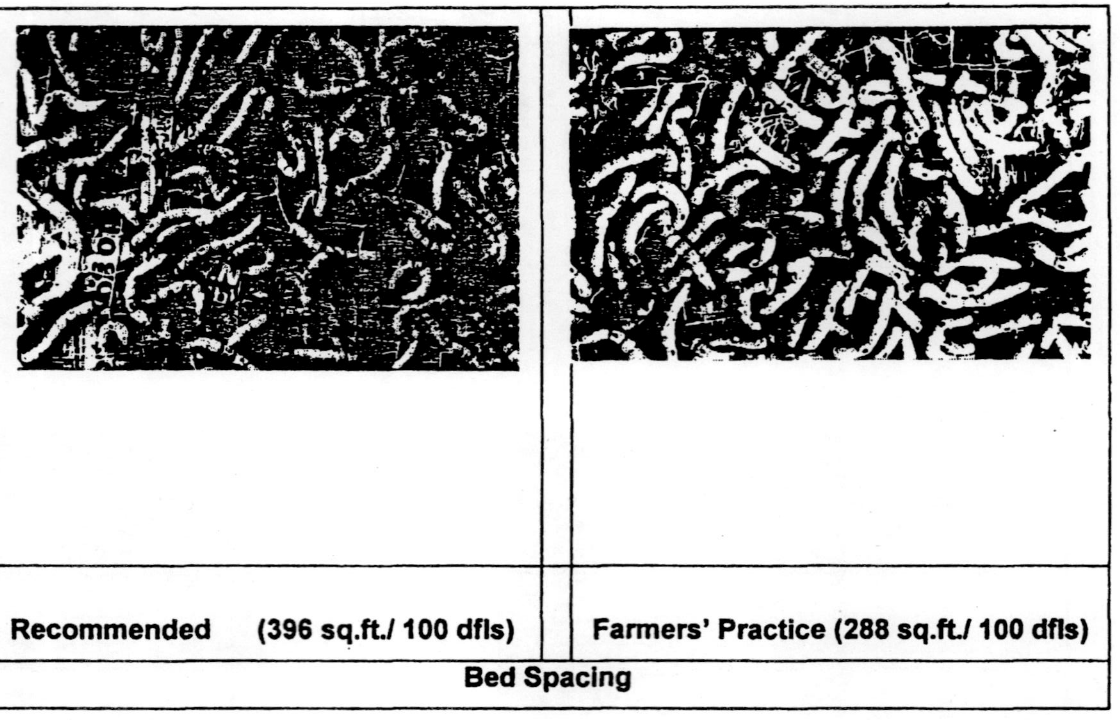

Fig. 17.3: Bed spacing (recommended and farmer's practice)

All treatments showed significant improvement in all the parameters tested over farmers' practice i.e. control.

TYPES OF MOUNTAGE

DISCUSSION

Providing optimum humidity is of paramount importance for the healthy growth of young age silkworm. If bed humidity falls below 70%, fast drying of mulberry leaves takes place which ultimately leads to loss of nutritive value of leaves. Thus, leaves become unsuitable for consumption and the larval growth slows down. Different methods of young age rearing are in practice. All methods aim to prevent drying of mulberry leaves fed by maintaining proper temperature and humidity in the rearing bed. Box rearing method performed best in dry summer and winter by providing the required microclimatic conditions for optimum growth and development of larvae leading to improvement of economic characters. Skuribayashi, (1995) also reported that in the box rearing method, mulberry leaves can remain fresh for a certain period of time. So the silkworms are well fed and mulberry leaf consumption as well as labour utilization becomes economical. This corroborates with the findings of Krishnaswami(1978) to achieve optimum relative humidity during chawki stage. For the same reason open type of shelf rearing registered its superiority during high humid seasons. Adequate spacing should be provided so that the silkworm develope in conformity with their full genetic potential. This findings corroborates with the finding of Sengupta & Yusuf, (1974) & Roychoudhury, 1990 & 1991 for both Chawki rearing and late age rearing, Iyenger, *et.al.*, (1988) pointed out that spacing schedule differs in respect of region, season and breed.

The quantity of food play an important role on growth and development of insects. The present investigation has given insight into the feasibility of providing leeser number of feeds to overcome the difficulties in high rainfall season in maintaining the regular four feeding schedules as also suggested by Guruswamy *et.al.*, (2002).

Table 17.5: Effect of mountage and mounting density on cocoon production during favourable season

Treatment	Coc. (%)	Shell (%)	Floss (%)	Fl.(m)	Renditta	Reelability (%)
T1	88	16.64	8.16	667	8.16	77.92
T2	86	16.37	8.60	662	8.30	74.81
T3	89	16.57	7.38	670	8.13	78.25
T4	89	16.34	8.26	650	8.28	77.82
T5	**92**	**17.02**	**5.73**	**719**	**7.91**	**80.84**
T6	90	16.83	6.05	709	8.21	79.47
Control	81	15.91	8.67	634	8.66	73.26
CD at 5%	**1.62**	**0.18**	**0.6**	**12.92**	**0.17**	**1.34**
CV%	**3.434**	**2.0189**	**14.703**	**3.566**	**3.799**	**3.209**

Table 17.6 Effect of mountage and mounting density on cocoon production during unfavourable season

Treatment	**Coc. (%)**	**Shell (%)**	**Floss (%)**	**Fl.(m)**	**Renditta**	**Reelability (%)**
T1	**91**	14.18	11.22	415	11.69	67.48
T2	89	14.01	11.26	406	11.71	65.29
T3	89	14.19	11.22	406	11.41	67.64
T4	89	14.06	11.81	411	11.64	66.53
T5	88	**14.29**	**8.95**	**448**	**11.03**	**74.13**
T6	88	14.23	9.10	437	11.25	70.60
Control	87	13.87	12.66	391	12.83	61.99
CD at 5%	**NS**	**0.11**	**0.68**	**11.91**	**0.28**	**1.87**
CV%	**3.996**	**1.456**	**11.577**	**5.285**	**4.488**	**5.095**

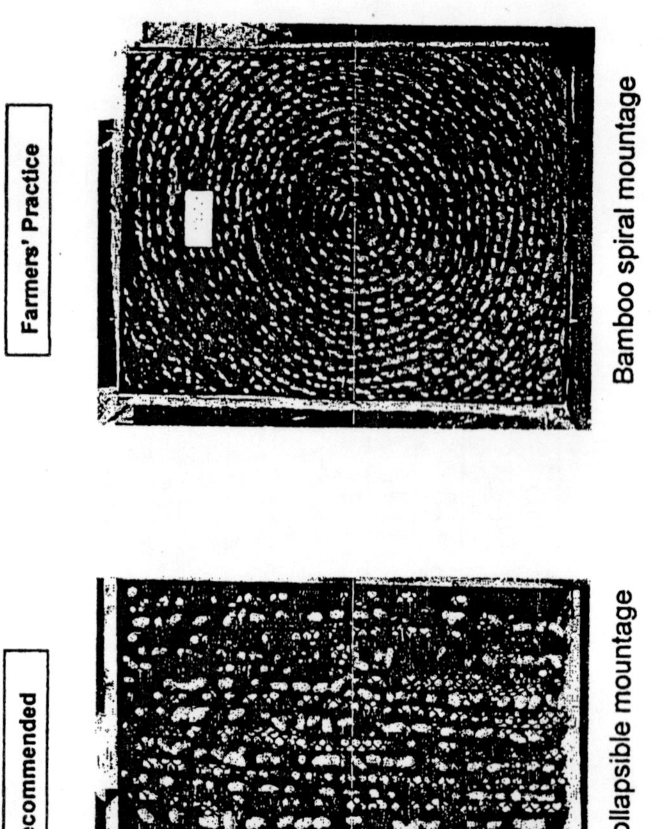

Fig. 17.4: Two types of mountage (recommended and farmer's practice)

Thus three feeding schedule per day with an interval of eight hour in between can effectively be undertaken which also enables minimum handling of the larvae and eliminates feeding of wet leaves during rainy seasons. Narayanan & Chawla, 1965 worked on feeding frequency and observed no significant difference in single cocoon wt, single shell wt, floss percentage, filament length and denier by practicing three times or four times or five times feeding in multivoltine silkworm breed. On the contrary, Hee, (1987) reported that larvae of *Bombyx mori* when fed 2 - 3 times daily have lower cocoon wt. compared to those fed four times daily. Oaquera, 1988 also suggested three feeding in the rainy season and four feeding during seasons with high temperature and low humidity (50-60%)

An analysis of cocoon crop data of silkworm *B. mori* clearly indicates that among the many factors that contribute to a good quality cocoon yield, the mounting material used for spinning of cocoon also play a vital role (Yokoyama, 1954, Tanka, 1964 & Alam *et al.*, 1977, Barah and Samson, 1990). In the present study, most of the cocoon characters and reeling parameters were found significantly superior in plastic collapsible mountage with 50 larvae /sq.ftfor Multi x Bi and 60 larvae /sq.ft for Multi x Multi. both in favourable and unfavourable seasons in comparison to Bamboo spiral mountage, Fig. 17.5.

It is evident that larval density during spinning plays important role in determining cocoon quality. Higher larval density can change the microclimate, which in turn affects the pupation (Reddy *et al.*, 2004). The present study also confirms this as the farmers' practice of mounting larvae in congested condition adversely affected both quality and quantity of cocoons.

Reddy *et al.*, (2004) also advocated that plastic mountages are more durable, easily washable and can be thoroughly disinfected, dried and preserve safely. It occupies less space and is suitable for self mounting and maintained cocoon quality. Geetha Devi *et al.*, 1990 opined that an ideal mountage having many advantages over conventional Chandraki is the plastic collapsible mountage. The present study also established the possibility of using plastic collapsible mountage by the farmers of Eastern India which will overcome the problem of scarcity of space during mounting and storage of mountage besides increasing cocoon quality.

CONCLUSION

Thus, inference can be drawn from this study that by adopting this season specific package of practices of silkworm rearing, rearers of Eastern India will be able to harvest qualitatively and quantitatively superior cocoon crops in all the commercial crop seasons inspite of fluctuating climatic condition. By utilizing this package of practices, a farmer can earn an additional cocoon yield of 9-10 kg and 11- 12 kg 7100 dfls during favourable and unfavourable crop seasons respectively which will fetch an additional income of Rs.900.00 -Rs.1000.00 (cocoon rate Rs.100/kg) and Rs. 880.00 - Rs.960.00 (cocoon rate Rs.90/kg) respectively. Considering the extra expenditure required , the net income comes around Rs.700.00 - Rs.800.00 for favourable season and Rs.680.00-Rs 760.00 for unfavourable season.

ACKNOWLEDGEMENT

The authors thank Mr. R. K. Das, TA, REC. .Mothabari and Mr. G.K.Mondal, T.A CSR&TI, Berhampore for technical assistance and data processing. Besides, help rendered by Mr. D. K. Mondal, TA and Md. M.Alam, TA is also highly acknowledged. The authors are also indebted to Mr. N.B.Kar, Scientist C, CSR&TI, Berhampore for Reeling data collection. Lastly, but not least, the help received from Time Scale Farm Workers of RTI Section, CSR&TI, Berhampore is also worth mentioning.

REFERENCES

Barah, A. & Samson, M.V. (1990). Effect of various mountages on the cocooning of the muga silkworm *Antheraea assama* Westwood. *Sericologia,* 30 (3): 313 - 321.

Geethadevi, R.G., Himanthraj, M.T., Vindhya, G.S. & Mathur, V.B. (1990). Can plastic collapsible mountage replace the bamboo mountage. *Indian Silk,* 29(6): 26 - 28.

Hee, B. (1987). *Sericulture Technology.* Seol National University, Press, pp. 52-144.

Iyanger, M.N.S., Nagaraj, C.S. & Reddy, S. (1988). Silkworm rearing practices in tropics. Lead paper. 3. *International congress of Tropical Sericulture practices*, Central Silk Board, India, 1-12.

Narayanan, E.S. & Chawla, S.S. (1965). Effect of frequency of feeding on the growth and development in two races of silkworm, local Mysore and Shungetsu Hosho. *Indian J. Seric.*, 4(4): 1-3.

Oaquera, H.S. (1988). The growth and cocoon yield performance of Silkworm multivoltine hybrid as affected by feeding frequencies. Dissertation submitted to CSR & TI, Mysore, p. 100.

Reddy, V.G., Venkatachalapathy, M., Manjula, A. & Kamble, C.K. (2004). Impact of mountages and larval density on cocoon quality and egg production. *Indian Silk,* Oct., 14-16.

Roychoudhury, N., Shamsuddin, M., Rao, G.S. & Kumar, T.P. (1990). Wider spacing improves bivoltine rearing. *News Views*, Jan.- June. pp. 7-8.

Roychoudhury, N., Paul, D.C. & Rao, G.S. (1991). Growth, fecundity and hatchability of eggs of *Bombyx moril.* in relation to rearing space. *Entomon.*, 16(3): 203-207.

Sengupta, K. & Yusuf, M.R. (1974). Studies on the effect of spacing during rearing on different larval and cocoon characters of some multivoltine breeds of silkworm *Bombyx mori.* L. *Indian J. Seric.,* 13: 11-16.

Tanaka, Y. (1964). *Sericology.* Translated and published by Central Silk Board, Bombay, India.

Yokoyama, T. (1954). *Synthesized Science of Sericulture.* Translated by CSB, Bombay, India, (1962).

NATIVE *BOMBYX MORI L.* (LEPIDOPTERA: BOMBYCIDAE): PROSPECTS OF IN-SITU MAINTENANCE

M.Z. Khan, K. Mandal, S.K. Das and A.K. Bajpai

ABSTRACT

A number of multivoltine silkworm breeds of *Bombyx mori* L. are available in various geographical zones of the country but their performance vary from their native place of origin which might be due to eco-friendly relationship with the particular environment. The quantum of intensity of biotic as well as genetic factors is also not certain, hence prospects of maintenance of silkworm in-situ draws a special attention as in other genetic resources. In the present paper, in-situ maintenance of various silkworm breeds/ races and the prospects of such measures to exploit their genetic potential have been discussed.

Key words: Nistari. Germplasm, Conservation.

INTRODUCTION

Indian sericulture is an ancient agro-based industry and reference to the use of silk in India is evident in ancient literatures like Vedas, 'Shankuntala' of Kalidasha (Watt, 1893). Although mankind maintains silkworm over centuries but information is scanty about *in-situ* maintenance of native *Bombyx mori* L. Sericulture in India was till recently confined in Kashmir and Uttar Pradesh in North, West Bengal and Assam in East, Karnataka, Andhra Pradesh and Tamilnadu in the South. Recently, under

Central Sericultural Research and Training Institute (CSR&TI), Berhampore-742101 (W.B.).
E-mails: drmzkhan1955@yahoo.com; akbjp@rediffmail.com

National Sericulture Project, sericulture has also been introduced in non-traditional states in India. Now, there is an urgent need to identify the silkworm races suitable for different agro-climatic conditions of various regions and seasons prevailing in India (Thangavelu, 1993).

In view of its domestication, the silkworm has become very sensitive and delicate and needs continuous care and attention. During the long history of sericulture it has been differentiated into a large number of distinct races with morphological and genetical differences.

The Insect Genetic Resources System (IGRS) advocates the conservation of genetic resources as one of the most important activities. Recent technological advancement has posed some practical problem to the native *Bombyx mori L.* due to their commercial value. The demand for hybrid seed at farmers level encourages the import of genetic material from the natural environment resulting in a 'stress situation' to the parent material to perform under vastly different conditions. The loss of genetic resources has been widely recognized and has received attention from scientists and conservationists (Rana *et al.*, 1993). The primary objective of germplasm preservation is to capture and maintain the germplasm variability of various species, to make this variability available to future generations of silkworm breeders. Hence, an attempt has been made here to discuss their maintenance in-situ, closely simultaneously their field performance, adaptability and survival, in the interest of further breeding programme in the country.

ECO-CLIMATIC CLASSIFICATION

India has been divided into four geographical zones on the basis of the climatic conditions i.e. temperate, sub-tropical, tropical humid and tropical dry. (Table 18.1). All the indigenous races of the country are distributed among these zones. All the indigenous are multivoltine silkworm races except rare uni/ bivoltine races in

India. The popular silkworm races like Nistari and Pure Mysore, now naturalized in India, were also introduced from China (Table 18.2).

Table 18.1: Geographical zones and sericultural states in India

Geographical zones	*Sericultural states*
Temperate	Kashmir.
Sub-tropical	Jammu, Uttar Pradesh, Himanchal Pradesh, Eastern States. North
Tropical humid	West Bengal, Orissa, Madhya Pradesh, Bihar.
Tropical dry	Karnataka, TamilNadu, Andhra Pradesh, Maharashtra.

Table 18.2: Indigenous and naturalized silkworm breeds in India.

Region	***Breed***	***Voltinism***	***Cocoon characteristics***
Bengal	Nistari	Multivoltine	Golden yellow, soft, flossy and spindle type cocoons. (Introduced from China in 1780-'81)
Bengal	Chotapolu	Multivoltine	Yellow or creamish white, small flossy cocoons.
Bengal	Barapolu	Univoltine	Greenish white, flossy one end pointed cocoons with short filament.
Assam	Sarupat	Multivoltine	Light creamy, small, flossy spindle shape cocoons.
Assam	Moria	Multivoltine	Creamy white, spindle shaped, small flossy cocoons.

Assam	Nayapolu	Multivoltine	Light yellow to white, flossy cocoon with short filament.
J&K	Jam	Bivoltine	White oval/constricted cocoons.
Karnataka	Pure Mysore	Multivoltine	Creamish yellow, flossy soft cocoons pointed at one end.
Karnataka	C. nichi	Multivoltine	White dumbbell shaped cocoons. (Introduced from China in 1875)
Karnataka	HS-6	Multivoltine	White dumbbell shaped cocoons.

NEED TO CONSERVE GENETIC RESOURCES FOR FUTURE:

Silkworm germplasm is irreplaceable and samples come in the forms of eggs, representing the genetic variation within the species; their wild relatives are currently being collected and conserved in gene banks and also in the field. The future of silkworm breeding efforts will depend on a continuing and expanding supply of genetic resources. The urgent work facing all users of silkworm germplasm is rescuing and preserving the available resources to meet the following needs:

I. To minimize the damage by major diseases and insects.

II. To control new pests.

III. To tolerate adverse environment.

IV. To improve quality and quantity.

V. To exploit the useful genes gathered from local races, evolved races and mutant races etc.

CONSERVATION STRATEGY

The best way to approach to the problem of conservation of genetic resources is to frame a conservation strategy. The choice of a conservation strategy depends on the nature of the material, their life cycle, mode of reproduction, size of the individuals and ecological status. The strategy, which aims to preserve as far as possible, the genetic identity of individual or population is termed as static conservation. The objectives of this conservation programme could be research, introduction, breeding etc. and this determines, the degree of suitability required to be maintained. The time scale determines the scope of the programme, concern for the preservation and the area or space to which it relates. For the silkworm races, this is meaningful in view of the native races; which are well-acclimatized and naturalized in particular areas and continually bringing new evolved races in their home environment i.e. *in-situ*. The *in-situ* conservation programme must be associated with sustainable forms of rural development in the seed zone areas.

Under this definition, in-situ conservation methods are those that maintain germplasm by paying due regard to the natural ecosystem of which the conserved population are a part. The natural biosphere is a useful solution for the species for the full expression of the genomes, whereby the full genetic potential of a particular race can be realised. However, conservation of eco-systems and biomass does not ensure continuing adoptive change unless genetic base of the species is sufficiently wide. Although silkworm has been domesticated long back, their conservation is still a challenging task due to various reasons. In every preservation practice the objective is to preserve at least one copy of different alleles. But, the genetic set up for the silkworm makes the minimum population size a formidable one. The problem is not simply confined to large scale and long duration preservation but also under the most viable conditions that they have to be maintained. Dynamic conservation can also be in freight (*ex-situ*) in the form of mass reservoir or gene banks and is useful for research for selection of locally adopted types. This has already been realized long back and establishments dealing with silkworm germplasm were maintained even in British India.

ENVIRONMENT AND SPECIES CONSERVATION

Each and every living organism has its specific surrounding medium or environment with which it continuously interacts and remains fully adapted. The environment is the sum total of physical and biotic conditions influencing the responses of the organisms (Kendeigh, 1974). The abiotic environmental factors can be classified mainly into two basic classes, namely:

I. Physical abiotic factors including temperature, humidity and spectral quality of light and darkness.

II. Chemical abiotic factors which include nutrient availability, pollutants present etc.

Temperature is one of the essential and obvious changeable environmental factors. It penetrates into every region of the biosphere and profoundly influences all forms of life by exerting its action through increasing or decreasing some of the vital activities of organism, like behaviour, metabolism, reproduction, ontogenetic development (viz., embryogenesis and blasotogenesis) and death. Temperature has a universal influence and is frequently a limiting factor for the growth or distribution of insects. Environmental temperature fluctuates both daily and seasonally. Different environments experience different temperature regimes. Temperature generally influences the behavioural pattern of insects and influences the completion of the several stages of their life cycle. Thus, temperature not only imposes a restriction on the distribution of species, but also their behaviour, metabolism, growth and development etc. After temperature, humidity is of prime concern. The existence of any insect species and its over all growth and development are dependent on humidity, if the humidity goes below a threshold limit, the viability of any species is in stake. The humidity and temperature can therefore, check and control any insect species.

Light and darkness, and the spectral quality have a bearing on insect population. More so, because some insect species are nocturnal, whereas some insect species are diurnal. The light intensity and spectral quality controls the insect population through their senses. Most of the insect species choose their suitable habitat on the basis of available photoperiods, luminescence and spectral quality. Some insects prefer some colours based on their typical colour preference.

FLUCTUATION OF POPULATION

The number of insect populations fluctuates in an irregular manner between very wide limits as a result of variability in the environment, supply of food and other resources. These fluctuate with time in localities where they are well established. The extent of these fluctuations varies in different years and in different places among different species. The apparent instability of numbers in such populations is characterized by the level of abundance about which numbers fluctuate, and this has led some workers to describe the determination of insect abundance mainly in terms of the intrinsic variability of environmental conditions. Population may influence numerical fluctuation to such an extent that their effects on one generation can be used to predict population numbers in the next, and is known as an equilibrium density (Morris, 1959). Dempster (1975) measured the factors that limit such temporal fluctuations and that restrict the spatial distribution by leading to extinction in some localities. Long-term age-specific life table data, gathered from a locality in which the impact of various factors, that affect mortality or natality, can be assessed.

STABILIZATION OF POPULATION

The species population is stabilized in the sense that they are prevented from increasing progressively with time by density-related subtractive processes. As a general working hypothesis, based on logical deduction and upon the empirical evidence obtained from recent field and laboratory studies, it can be assumed that population persistence depends upon:

I. The influence of the environmental factors as well as the species characteristics, this in turn contributes towards setting the ranges between which numbers may fluctuate.

II. The operation of processes, related to population density by negative 'feed-back' act as stabilizing mechanisms to limit numerical increase. Hence, it is essential that the population density and persistence of a race in a particular environment may be assessed and its fitness expressed as a function of its prosperity to multiply.

FUNCTION OF LIFE SYSTEMS

Insect species differ greatly in the characteristics in their living environment. The insect number can serve as a guide for the study of particular populations (Solomon, 1949 and 1957; Andrewartha and Birch 1954; Chitty 1960 and Huffaker and Messenger 1964). Now, it is possible to propose a way of generalizing the determination of insect numbers on the basis of the results of recent population studies. A comprehensive synthesis can be achieved by:

I. 'Life system' that incorporates all influences which may affect population numbers; and

II. The interactions of general characteristics of life systems and complex webs that determine the persistence and abundance of populations. Population generally restrict within an eco-system and within circumscribed limit due to interaction of other populations and the physical conditions in the nature (Bakker 1964). The life system is composed of species and its effective environment. Life system determines the existence of a particular population, which is a part of an ecosystem (Fig.18.1):

DISCUSSION

It has already been established that the genotypic and phenotypic factors are responsible for the expression of the genome of any living organisms. Their certainty and quantum has not been established so far. As India is a vast country, naturally, it exhibits a great variety of vegetation. The panorama of geographical variations and distinctions, which range from the rain forests of

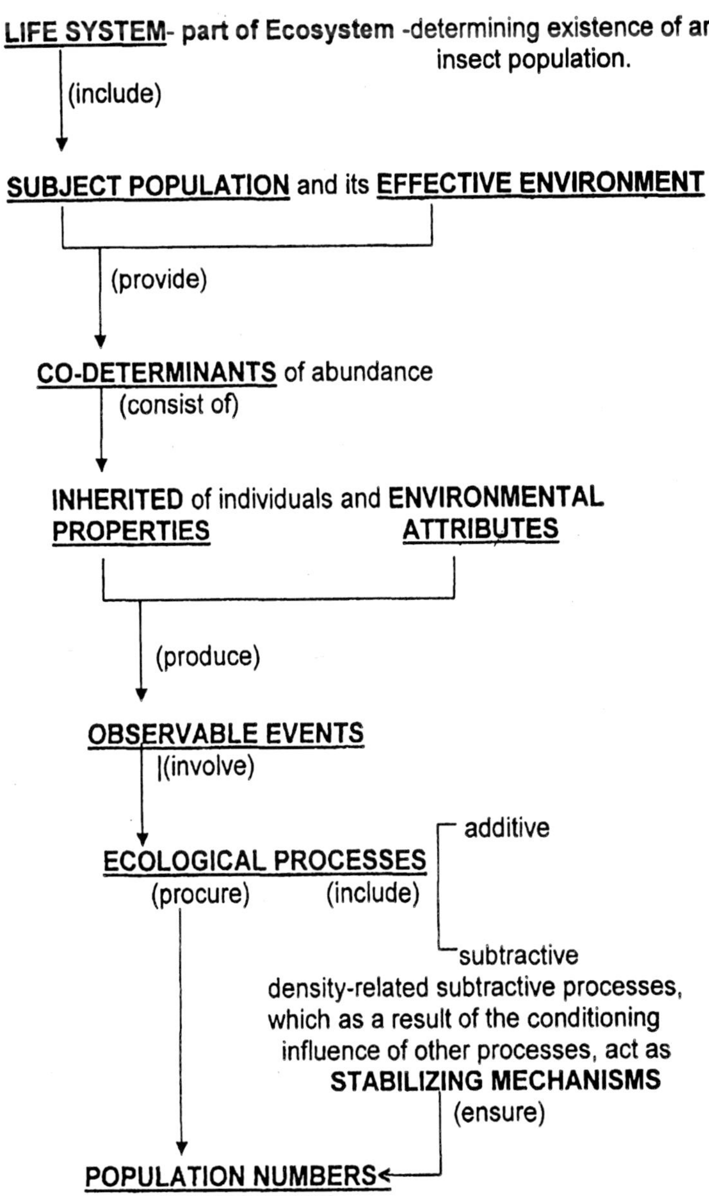

Fig. 18.1: The element involved in the functioning of a life system.

Assam to the snows of the Himalayas and the deserts of Rajasthan, from deciduous forests of central highlands to mangrove swamps of the Sunderban, have their own typical flora and fauna assemblage. These flora and fauna complexes have played vital role in the development of human culture from the dawn of civilization. Man by their typical interaction has maintained or destroyed organisms; and many races have become extinct for such interactions. Bulupolu and Kasmiri races have been extinct due to one or another reasons. The Nayapolu, Chotapolu may also vanish tomorrow unless the same type of the protection measures necessary for the threatened animals in situ (viz., Project Tiger, Gir Lion Sanctuary Project, Crocodile Breeding Project, Himalayan Musk Deer Project, Manipur Brow antlered Project, Lesser Cat Project) are taken up. Our honourable Prime Minister, Smt. Indira Gandhi very rightly highlighted it "A threat to any species of plant and animal life is a threat to man himself. As such, it is essential that we should open our minds right now and think about the vast potential and to conserve them *in-situ.*

The National Bureau of Plant Genetic Resources has already setup various germplasm conservation stations being eco-friendly relationship of various genetic resources all over India realizing the importance of *in-situ* maintenance. Due to lack of the prime importance of germplasm, many third world countries now face the threat of genetic erosion in their genetic store. Varieties of germplasm are becoming extinct on the heels of development and the nation is paying an exorbitant price in the name of exotic varieties. In India the awareness about the role of genetic resources in sericulture was initially belated. Now, Central Silk Board has established various institutes as well as research stations all over India, particularly in West Bengal, Assam, J&K and Karnataka where the indigenous silkworm races have been naturalized. Only the need is to identify the particular and specific race of any specific area through strict survey and to maintain the same *in-situ*. It can be stated that single nuclear germplasm bank located at one particular place cannot serve all the needs, rather in-situ conservation at their place of origin is highly desirable.

The benefit of *in-situ* conservation is manifold. Firstly, the environmental acclimatization helps the species to overcome slight environmental disturbances-fluctuation in temperature and humidity. Secondly, in their home, they are more at ease over pathological enemies, predators, and parasites. Thirdly, *in-situ* presupposes local technique in rearing that has passed over time and found beneficial in final score. Fourthly, a permanent zonal occupancy fixes limits on predators and pests through spatio-temporal blocked rearings, existence of hyper parasites etc. However, for the safety purpose in-freight (*ex-situ*) parallel stock can also be maintained to supplement in case of major population losses.

ACKNOWLEDGEMENT

Grateful thanks to Mr. Indrajit Ray, Central Sericultural Research and Training Institute, Berhampore for the preparation of the manuscript.

REFERENCES

Andrewartha, H.G. & Birch, L.C. (1954). *The distribution and abundance of animals*. Chicago: University of Chicago Press. Chitty, D. (1960).

Population processes in the vole and their relevance to general theory. *Can. J.* Zool., 38: 99-113.

Dempster, J.P. (1975). *Animal Population Ecology*. Academic, London. Huffaker, C.B. & Messenger, P.S. (1964). The concept and significance of natural control., pp.74-117.

Debach, P. [ed.] *Biological control of insect pests and weeds*. London: Chapman and Hall. Morris, R.F. (1959). Single factor analysis in population dynamics. *Ecology*, 40:580-588.

Rana, R.S., Gupta, P.N. & Rai, M. (1993). Genetic resources programme of horticulture crops. Indian Prospective: *1-Proceeding of Golden Jubilee Symposium, Horticulture Research- Changing Scenario.*

Solomon, M.E. (1949). The natural control of animal populations. *J. Anim. Ecol.*, 18: 1-35.

Solomon, M.E. (1964). Analysis of processes involved in the natural control of insects, In *Advances in ecological research* (ed. Cragg, J.B.). Vol. 2. London and New York, Academic press, pp. 1-58.

Thangavelu, K. (1993). Lacunae in Indian sericulture. *Indian Seri.*, 32(4): 13-19.

Watt, G. (1893). *A dictionary of the economic products of India.* Vol. VI (3):25. Cosmos Publications, Delhi.

INTRODUCTION

Being a bivoltine cocoon producing state, H. P. is taking two silkworm crops in a year i.e. spring and autumn. The spring crop is main with 65% of annual rearing capacity while autumn only 35% rearing is in practice. Being suitable climatic condition in spring crop, the incidence of silkworm diseases is less i.e. 15% to 20% while it is on the higher side in autumn crop i.e. 30% to 35%. The crop loss due to silkworm diseases like grasserie and flacherie is main cause of less sustainability of sericulture in Himachal Pradesh. Though, the farmers are using bed disinfectant produce no desired results to control the silkworm diseases and increase the cocoon productivity. To overcome this problem an experiment was designed by REC, Una to use the different bed disinfectant i.e. Vijetha and RKO in both the crops as per their use of recommendation.

MATERIALS AND METHODS

15 silkworm rearers were selected for this study, out of which 5 farmers were dusted vijetha, 5 farmers were dusted RKO and remaining 5 farmers were kept as control using no disinfectant during late age rearing. All the groups reared 250 dfls during spring 2006 while 200 dfls per group during autumn 2006. Other rearing conditions were kept same in each group. The leaf quality fed to the silkworm was also from the same type of trees/mulberry varieties. Prior to the commencement of the rearing, all the rearing houses and other material used in rearing were disinfected with serichlor. The rearing was conducted on shelves and the shoot feeding was given to silkworms. The farmers were supplied the bed disinfectants @ 2.850 kg per 100 dfls for late age rearing and were asked to apply the same on the basis of recommendation i.e. 550 gms after 3rd moult, 800 gms after 4th moult and 1.500 kg on the 4th day in 5th instar. The bed disinfectant was applied in presence of technical staff of REC. The silkworm were fed with tender leaf after half an hour of application of bed disinfectant at the time of moult resumptions. The larval period was also same

Table 19.1: Comparative results of use of bed disinfectants against control during spring crop 2006.

S. No.	Name of the farmer	Village	Dfls reared (ozs.)	Bed disinfectant used	Total yield (kgs.)	Yield/oz (kg)	Increase in productivity (%)
1	Sh. Kashmiri Lal	Nangal Sal	0.50	Control	25.500	51.000	——
2	Smt. Shanti Devi	Surjehra	0.50		25.200	50.400	
3	Sh. Aman Vashishta	Charatgarh	0.50		24.000	48.000	
4	Smt. Naresh Kumari	Charatgarh	0.50		24.800	49.600	
5	Smt. Santosh Kumari	Takka	0.50		25.6	51.200	
			2.50		**125.100**	**50.040**	
6	Smt. Beena Devi	Takka	0.50	**Dusting of RKO**	33.000	66.000	**14.78**
7	Sh, Brij Mohan	Takka	0.50		26.800	53.600	
8	Sh. Suresh Sharma	Surjehra	0.50		28.300	56.600	
9	Smt. Ram Asri	Jhember	0.50		27.800	55.600	
10	Sh. Shyam Lal	Basal	0.50		27.700	55.400	
			2.50		**143.600**	**57.440**	
11	Smt. Bindu Sharma	Takka	0.50	**Dusting of Vijetha**	30.800	61.600	**23.90**
12	Smt. Sheela Devi	Badoli	0.50		27.600	55.200	
13	Smt. Sulochna Devi	Takka	0.50		33.800	67.000	
14	Smt. Sudesh Kumari	Basal	0.50		33.800	67.600	
15	Smt. Asha Rani	Una	0.50		29.000	58.000	
			2.50		**155.000**	**62.000**	

Table 19.2: Comparative results of use of bed disinfectants against control during autumn crop 2006.

S. No.	Name of the farmer	Village	Dfls reared (ozs.)	Bed disinfectant used	Total yield (kgs.)	Yield/oz (kg)	Increase in productivity (%)
1	Sh. Jagmohan	Badehia	0.50	**Control**	17.000	34.200	
2	Sh. Rajendra	Joh	0.50		17.900	35.800	
3	Sh. Amar Nath	Basal	0.50		17.500	35.000	
4	Sh. Rakesh	Bhadsali	0.25		8.500	34.000	
5	Sh. Sanjeev Kumar	Dioli	0.25		8.000	32.000	-
			2.00		**69.000**	**34.500**	
6	Sh. Kuldeep	Ambota	0.50	**Dusting of R.K.O.**	19.000	38.000	
7	Smt. Saroj	Andora	0.50		20.100	40.200	**10.58**
8	Sh. Darshan	Abheyapu	0.50		18.400	36.800	
9	Sh. Balbir Singh	Kuthar	0.25		9.500	38.000	
10	Sh. Mahender	Ambota	0.25		9.300	37.200	
			2.00		**76.300**	**38.150**	
11	Smt. Gurmeto	Badoli	0.50	**Dusting of Vijetha**	20.300	40.600	
12	Sh. Ramesh	Nari	0.50		21.600	43.200	
13	Smt. Ram Asri	Batuhi	0.50		20.100	40.200	**22.3**
14	Sh. Mehar Cha.	Batuhi	0.25		11.500	46.000	
15	Smt. Rami Devi	Lower Badehra	0.25		10.900	43.600	
			2.00		**84.400**	**42.200**	

in all the batches of experimental lots. The mounting material used was also same to all the groups.

RESULTS AND DISCUSSION:

The study reveals that bed disinfectants have a vital impact to control the silkworm diseases and enhance the cocoon yield in both the crops. Vijetha proved better bed disinfectant than RKO in both the crop. The results are given in Table 19.1 and 19.2. Though bed disinfectant proved effective in both the crops but higher cocoon yield in spring crop is due to congenial climatic condition and better quality of leaf, which was not available in autumn crop. Even then bed disinfectant in both the crops maintain hygienic condition in the bed which prevented multiplication of pathogens resulted less incidence of Silkworm diseases especially in autumn crop. Vijetha proved better than RKO which seems to be its constituents used in its manufacturing.

CONCLUSION

After analyzing the results it is clear that use of bed disinfectant by the rearers during late age silkworm rearing is must to control the silkworm diseases and increase cocoon productivity. The use of bed disinfectant is cost effective as the revenue generated through cocoon production is more which compensates the cost of bed disinfectant by giving high yield. The study proved that vijetha is most effective bed disinfectant, which should be polularized atleast in autumn crop which will help in prevention of fatal disease like grasserie and flacherie.

REFERENCE

Tayal, M.K., Chauhān, T.P.S., Tiwari, P., Mishra, P. N. & Khan, M. A. (2007): Uttar Paschim Bharat Main Safal Keetpalan Hetu Akikrit Resham Keet Rog Prabhandhan Ka Mahatava Avam Awashayakta. *Akhil Bhartiya Rajbhasha Taknik Seminar*, pp. 102 – 106.

INDEX